Your Digital Afterlives

Palgrave Frontiers in Philosophy of Religion

Series Editors: **Yujin Nagasawa and Erik Wielenberg**

Titles include:

Zain Ali
FAITH, PHILOSOPHY AND THE REFLECTIVE MUSLIM

István Aranyosi
GOD, MIND AND LOGICAL SPACE
A Revisionary Approach to Divinity

Yujin Nagasawa (*editor*)
SCIENTIFIC APPROACHES TO THE PHILOSOPHY OF RELIGION

Aaron Rizzieri
PRAGMATIC ENCROACHMENT, RELIGIOUS BELIEF AND PRACTICE

Eric Charles Steinhart
YOUR DIGITAL AFTERLIVES
Computational Theories of Life after Death

Forthcoming titles:

Trent Dougherty
THE PROBLEM OF ANIMAL PAIN
A Theodicy for All Creatures Great and Small

Aaron Smith
THINKING ABOUT RELIGION
Extending the Cognitive Science of Religion

Palgrave Frontiers in Philosophy of Religion
Series Standing Order ISBN 978–0–230–35443–2 Hardback
(*outside North America only*)

You can receive future titles in this series as they are published by placing a standing order. Please contact your bookseller or, in case of difficulty, write to us at the address below with your name and address, the title of the series and the ISBN quoted above.

Customer Services Department, Macmillan Distribution Ltd, Houndmills, Basingstoke, Hampshire RG21 6XS, England

Your Digital Afterlives
Computational Theories of Life after Death

Eric Charles Steinhart

Department of Philosophy, William Paterson University of New Jersey, USA

First published 2014 by
PALGRAVE MACMILLAN

Palgrave Macmillan in the UK is an imprint of Macmillan Publishers Limited, registered in England, company number 785998, of Houndmills, Basingstoke, Hampshire RG21 6XS.

Palgrave Macmillan in the US is a division of St Martin's Press LLC, 175 Fifth Avenue, New York, NY 10010.

Palgrave Macmillan is the global academic imprint of the above companies and has companies and representatives throughout the world.

Palgrave® and Macmillan® are registered trademarks in the United States, the United Kingdom, Europe and other countries.

ISBN 978–1–137–36385–5

This book is printed on paper suitable for recycling and made from fully managed and sustained forest sources. Logging, pulping and manufacturing processes are expected to conform to the environmental regulations of the country of origin.

A catalogue record for this book is available from the British Library.

A catalog record for this book is available from the Library of Congress.

Transferred to Digital Printing in 2014

Contents

Figures and Tables

Figures

Tables

Series Editors' Preface

The philosophy of religion has experienced a welcome revitalization over the last 50 years or so and is now thriving. Our hope with the *Palgrave Frontiers in Philosophy of Religion* series is to contribute to the continued vitality of the philosophy of religion by producing works that truly break new ground in the field.

Accordingly, each book in this series advances some debate in the philosophy of religion by offering a novel argument to establish a strikingly original thesis or approaching an ongoing dispute from a radically new point of view. Each book accomplishes this by utilizing recent developments in empirical sciences or cutting-edge research in foundational areas of philosophy, or by adopting historically neglected approaches.

We expect the series to enrich debates within the philosophy of religion both by expanding the range of positions and arguments on offer and by establishing important links between the philosophy of religion and other fields, including not only other areas of philosophy but the empirical sciences as well.

Our ultimate aim, then, is to produce a series of exciting books that explore and expand the frontiers of the philosophy of religion and connect it with other areas of inquiry. We are grateful to Palgrave Macmillan for taking on this project as well as to the authors of the books in the series.

Yujin Nagasawa
Erik J. Wielenberg

Preface and Acknowledgments

All will agree that computers have radically changed our ways of living – but they have also changed our ways of thinking. One of the more surprising consequences of the computer revolution is that our digital technologies have provided us with new and more naturalistic ways of thinking about old religious topics. *Digitalism* is a philosophical strategy that uses these new computational ways of thinking to develop naturalistic but meaningful approaches to religious problems involving minds, souls, life after death, and the divine. *Your Digital Afterlives* develops digitalism.

My greatest source of encouragement for the development of digitalism has been Jim Moor, in the Philosophy Department at Dartmouth College. He encouraged me to develop many articles dealing with digitalist ideas. John Leslie deserves a great deal of credit for keeping me going through this project. He is the source of many conceptual breakthroughs that made my own thought possible. Jack Copeland merits many thanks for helping me with my work on infinite minds and for sharpening my thinking about digital gods. Nick Bostrom provided encouragement for my research on the theological implications of his simulation argument. I would also like to thank Graham Priest for his support.

I gave a paper at the Pacific APA in San Francisco in April 2007 linking John Hick's resurrection theory to temporal counterpart theory. David Vander Laan gave me very helpful feedback. Peter Byrne helped me with my earlier work on the revision theory of resurrection, which inspired much of my later work on life after death. John Hick himself encouraged earlier versions of *Your Digital Afterlives* and I especially thank him.

Robin Le Poidevin has also given me much-needed support during my work on the plurality of gods. I also thank Klaas Kraay for his insistence that I clarify those ideas. His work on the theistic multiverse inspired me. Over the course of several workshops, Hugh Woodin helped me think more clearly about infinity and the divine. At those same workshops, Wolfgang Achtner gave me much good theological advice. My involvement in those workshops was graciously funded by the John Templeton Foundation, despite the clearly advertised fact that my own theological orientation is far from conventional. Bruce Reichenbach and John

Larry Crocker gave me some very valuable criticisms. Dan Fincke and Pete Mandik constantly challenged me with objections. Without their criticisms, these ideas would be much weaker. Yujin Nagasawa deserves enormous credit for his vision of a more diverse and more intense future for the philosophy of religion. Brendan George, at Palgrave, has been a wonderful editor. And Melanie Blair, also at Palgrave, has provided invaluable assistance.

I doubt that I'd be able to do any philosophical work at all without the constant support of my wife, Kathleen Wallace. I also thank Jed Williamson and Perry Forbes Williamson for the many summers we spent at Camp Everhappy. I greatly appreciate the support of Dartmouth College and William Paterson University.

I
Ghosts

1. Digital ghosts

After you die, you can *remain* in something else. You can and do remain in any thing that carries information about your life. Remaining is a matter of degree: you remain more in things that carry more information about your life. You can remain in other living things. If you have any offspring, then some of your genes remain in their genes. And aspects of your life can remain in the memories of others. Your ideas and values can remain in the lives of other people. But you can also remain in things that do not live. You can remain in your skeleton, in your mummy, in your preserved DNA (deoxyribonucleic acid).[1]

You can remain in artifacts. You can remain, like Samuel Pepys, in a diary written in ink on paper. Pepys compiled a detailed diary entry for every day of his life from January 1660 to May 1669. Almost 400 years later, those days of his life remain, at least partly and approximately, in those diaries. You can remain in photographs or videos or recordings of your voice. Obviously enough, these are becoming more and more common, and they can be looked at or listened to by others after you die. Sadly, those artifacts record only your superficial features – and they are not interactive. Fortunately, digital technologies are enabling us to produce interactive digital diaries (Blascovitch & Bailenson, 2011: ch. 9). You can remain in (or *as*) an interactive digital diary.

An interactive digital diary is a *digital ghost*.[2] As digital technology makes progress, our digital ghosts are becoming more and more accurate descriptions of our lives. They are recording more aspects of our lives at ever greater levels of detail. And our digital ghosts are slowly becoming more interactive – they are slowly becoming intelligent. Any possible future digital ghost carries information about your life. Since

1

it carries that information, you remain, after death, in your ghost. Of course, digital ghosts are not like the ghosts described by spiritualists or occultists. Digital ghosts are entirely physical patterns of energy in material computing machines. They are not supernatural.

A good way to illustrate the concept of your digital ghost is to present several generations of technically possible ghosts (Steinhart, 2007a). The first generation includes the ghosts that exist in the early twenty-first century. These ghosts are primitive – really, they are merely proto-ghosts. The next generations include various improved types of digital ghosts. At least in some possible human futures, as technology makes progress, *perfect* digital ghosts will eventually become available. However, bear in mind that the series of generations of ghosts is not intended to serve any prophetic purposes. It merely describes various stages in the *conceptual* evolution of digital ghosts.

2. Your Facebook timeline

Many people in the early twenty-first century are building digital ghosts.[3] For the sake of clarity, it will be helpful to focus on one type of first-generation ghost. Your first-generation ghost is your *Facebook timeline* (Abram, 2012: ch. 4; Harvell, 2012: ch. 7). Your Facebook timeline is a temporally ordered sequence of posts. Each post describes some aspect of your life at some time – it describes some *stage* of your life. Your posts might be detailed descriptions of your entire day (like the entries in the diary of Pepys). Or they might be descriptions of shorter events. More technically, your Facebook timeline is a temporal database, and your posts are records entered into that database.

Many different kinds of digital records can be posted to Facebook timelines. These records include written statements, photographs, and video and audio files. They also include lists of your friends (your social network). Data that can be recorded on a Facebook timeline also includes various personal facts (such as your relationship status, places of education and employment, and so on). Facebook timelines also include records of various activities performed on Facebook, such as acts indicating that you *like* an item on Facebook, as well as comments you post on the timelines of your friends.

After you die, Facebook currently creates a memorialized version of your timeline.[4] Your memorial timeline preserves most of the content of your original timeline. Currently, after you die, it can only be accessed by your Facebook friends. If one of your friends reads your memorial timeline, then they are *visiting* your digital ghost. Your visitors can scroll through your timeline to find records of earlier parts of your life. Obviously, your first-generation ghost isn't very interactive. Programs exist that can infer many personal attributes from your Facebook activity (Kosinski et al., 2013). For example, they can infer your political and sexual preferences from your patterns of liking. If someone could run those programs on your memorial timeline, then they might learn many new things about you. However, those programs are not presently available to the general public.

3. Your totally quantified self

Your second-generation digital ghost is initialized at birth.[5] The first record entered into your new ghost contains your basic biological data (for instance, your genetic code, blood type, fingerprints, an image of your iris, and so on). When you are young, your parents manage your ghost. They fill it with data. As you grow older, you take more responsibility for entering your data into your ghost yourself. This second-generation data includes all the data that can be entered into your first-generation ghost. But most of your second-generation data is entered into your ghost automatically.

Your second generation data includes many of your perceptual inputs. Your visual and auditory inputs are constantly recorded by tiny cameras and microphones (such as those in Google Glass). Your second-generation data includes many physiological descriptors (such as your heart rate, blood pressure, skin conductance, body temperature, blood glucose levels, breath chemistry, and so on). It includes the electrical activities of your heart and brain – your electrocardiogram and electroencephalogram are constantly recorded. All your medical records are posted to your ghost (including the results of physical and psychological exams).[6] Your second-generation data includes many of your behavioral outputs (the motions of your limbs are recorded). Anything you say or type is recorded.

As data enters your digital ghost, it serves as the raw material for the synthesis of a daily model of your stimulus–response patterns. Your second-generation ghost consists of a series of these daily ghost patterns, each of which approximates your psychology on some given day. Each daily ghost pattern is an artificially intelligent program. From your biographical data, it endeavors to reconstruct what you experienced, how you felt, and what you did. From the records of your perceptual inputs and behavioral outputs, it strives to infer what you thought and what you desired. It tries to figure out your preferences, habits, values, virtues, and vices. On the basis of careful study of your biographical data, each daily ghost is an expert on your life up to that day. And, through that expertise, its artificial personality imitates your organic personality. It is your digital surrogate.

After you die, your ghost remains – it is an enormous system of files stored on some digital media. Of course, your ghost is entirely physical. Every part of your ghost is some pattern of electromagnetic energy, stored on some physical substrate. From time to time, your ghost may receive visitors. When a visitor wants to interact with your ghost, they select a specific day of your life. Your ghost mind for that day is loaded onto a computer, which animates it. The computer that runs your second-generation ghost is a very powerful version of the computers we currently have on our desks. More technically, it is a *von Neumann machine*. This computer produces visible outputs on some screen. When your visitor looks at this screen, they see your ghost face. Your ghost face looks like your face on that day, and it moves like your face on that day.

Your ghost can hear through microphones attached to its computer, and it can talk through loudspeakers. Your ghost can carry out a conversation just as well as you could carry out a conversation on the selected day. Your ghost is an *embodied conversational agent*, also known as a *chat bot* (Cassell et al., 2000). So, if your ghost is old enough to talk, the easiest way to interact with it is verbally. Your visitor asks questions, and your ghost answers pretty much like you would have answered. When your ghost talks, its voice sounds like your voice sounded on that day. Your ghost can answer factual questions about your life up to the selected day; it can explain your behaviors; it can express your views and opinions. Your ghost can produce many other forms of output besides speech: it can produce text, photos, videos, and audio. Still, your ghost is far from having anything like full personal intelligence. Its conversations are factual and dull.

4. Your ghost brains day by day

As before, your third-generation ghost is initialized at your birth. It includes all the data that gets entered into your second-generation ghost. But your third-generation ghost gets data from several new sources. Specifically, it gets data from embedded sensors and from detailed movies of your brain at work. These brain movies are produced by very high definition versions of functional magnetic resonance imaging or related scanning technologies.

As data enters your third-generation ghost, it serves as the raw material for the synthesis of a daily model of your brain. Your digital ghost now consists of a sequence of these daily ghost brains, each of which approximates your brain on some given day. Your ghost brains structurally and functionally resemble your brain. Each ghost brain is an artificial neural network – it is a big connect-the-dots structure in which the dots are software neurons and the connections are software synapses. Still, your ghost brains are not fully accurate models of your brain. They are functionally simplified versions of organic brains. And your ghostly neural networks are simplified versions of organic neural nets. Your ghost brains do not replicate your brain at the levels of synaptic or molecular detail.

After you die, your ghost brains are just files stored in some big database. As before, when a visitor wants to interact with your ghost, they select a specific day of your life. Your ghost brain for that day is loaded onto some computer. Perhaps this computer is a vast network of von Neumann machines (with more power than the entire Internet, but in a box on your desk). Perhaps this computer is a network of neuromorphic chips, specially designed to simulate neural circuitry. This computer animates your ghost brains. As it animates those brains, it uses their memories to reconstruct the environment in which you found yourself on that selected day. The memories in your ghost brain are used to build a virtual reality model of the world as you experienced it on that day. This virtual reality model of your experienced environment is your *ghost world*. It includes many *ghostly objects* – digital reconstructions of things you experienced. And since you always experienced your body, your ghost world contains a *ghost body*.

Of course, your ghost brain, your ghost body, and all the objects in your ghost world are realized by patterns of energy in some physical computer. If the computer that realizes your ghost and its world is an electronic machine, then your ghost world is an electronic world and the objects in it are electronic objects. As such, they are physical. More precisely, since electrons are material things, your ghost body and all the objects in your ghost world are material things. Nevertheless, they are not *ultimately real things* – they are realized by the deeper physics of the computer which brings them into existence. Since they are not ultimately real, we will say that they are *virtual*. But virtual things are not fake. And they are certainly not abstract objects like numbers. Virtual things are physical things which continuously depend on other physical things. Virtual things are software things while ultimately real things are hardware things. Your ghost world, including your ghost body, is a software process running on some hardware substratum.

When a visitor summons your ghost on a specified day, your ghost replays its record of that day. Your visitor can watch, passively, as your ghost goes about its business. This is like watching a movie. The script for the movie is the biographical record compiled during your life on that day. And this script is pretty accurate. Your visitor can watch this movie from a variety of points of view, but they all have to be pretty close to your original point of view – your visitor sees your original world through your ghostly eyes. If your daily routine seems boring, your visitor can hit fast-forward. Your ghost then replays your biography at a higher speed. Your visitor can jump forwards or backwards, skipping over parts of your life. Your visitor can search for specific activities, people, places, or other things, and can tell your ghost to play the relevant scenes.

But your visitor can do far more than just watch your ghost. Since your ghost now replicates your intelligence at a very high level of accuracy, it can also hold conversations. Your visitor can converse *intelligently* with your ghost. Your visitor can ask your ghost what it's doing, why it's doing it, and so on. Your visitor can ask your ghost about people or places or things in your ghostly environment. It can ask any question at all. Your ghost always gives an answer that is based as closely as possible on its memories and on any other biographical data collected for that day. If your ghost can answer quickly, it keeps going about its business. Otherwise, it pauses to figure out its answer or to give a longer speech. During that pause, its ghost world pauses too. Once it's done, it starts going about its business again. A visitor can keep your ghost paused for as long as they like, asking it all sorts of questions, and getting its answers.

But how does your visitor appear to your ghost? Your visitor appears as a disembodied voice, like the voice of an invisible spirit sitting on its shoulder. For your ghost, of course, talking with an imaginary friend is normal and natural. It doesn't upset your ghost or make it feel ill. Your ghost brain has been programmed to politely accept such interactions. And perhaps an advanced third-generation visitor might control their own ghostly hand, so that they can point to something and ask your ghost about it. Still, your visitor does not have much of a presence in your ghost world – they don't have their own virtual body. They don't have an avatar which your ghost can see and touch (and vice versa). And your visitor has very little control over your ghost, since there isn't much that can be controlled. All your ghost can do is to replay your life and answer questions about it.

Visitors are not really needed for the reanimation of your ghost brains. Somebody can select some period of your life (such as your first ten years) and instruct the computer to reactivate your ghost brains for that period. For any series of days, the computer starts with the ghost brain on the first day, and plays brain after brain, day after day. It uses the series of daily ghost brains as landmarks for the synthesis of a psychologically continuous brain-process that runs across many days. The life of your brain can be replayed in the computer whether or not any visitor is watching. If this is done, then a replica of your brain-process will exist in a computer after your death. Of course, the computer that replays your brain-process might be much faster than your organic brain. It might be able to reproduce the brain-process of your entire organic life in a matter of minutes or seconds. And this would be a very thin kind of life after death – it would be your digital afterlife.

5. Replaying your mental life

Your fourth-generation ghost is an enhanced version of your third-generation ghost. It consists of a series of ghost brains. But now these ghost brains are recorded at much higher levels of biological fidelity. They are connect-the-dots networks in which each dot accurately imitates one of your neurons and each connection accurately imitates one of your synapses. Your ghost brains now replicate the functionality of your brain at the molecular level. And your ghost brains are recorded very frequently. Perhaps your fourth-generation ghost contains a ghost brain for every millisecond of your life.

After your death, your ghost brains are stored in digital files. When a visitor wants to interact with your ghost, they select a specific period of your life. Your ghost brains for that period are loaded into a computer. When your visitor launches your ghost, the computer begins to simulate your brain-processes for that period. By replaying your brain-processes, the computer constructs your ghost world from your memories. As before, your visitor can appear to your ghost as a disembodied voice and hand. But the fourth generation of ghosts adds something new. At least the perceptual inputs to your ghost brain can be transmitted directly to the relevant circuits in the brain of your visitor. For example, the signals in your ghostly optic nerves can be transmitted directly to the optic nerves of your visitor. Thus your visitor can visually experience your world from your first-person perspective. Your visitor can see exactly what you saw exactly as you saw it.

Of course, many philosophers argue that any simulation of your brain is also a simulation of your mind. They say that mental processes are identical with brain-processes. And, since brain-processes are computable, mental processes are computable too. The *classical digitalists* include writers like Edward Fredkin, Hans Moravec, and Ray Kurzweil. They all argue that mentality is entirely computable.[7] Following the classical digitalists, *all* digitalists agree that mentality is entirely computable. Minds are software processes running on hardware substrates – minds are computations. Therefore any mind, including yours, can be exactly replicated in a computer. Digitalists are not *mysterians* – they resolutely refuse to mystify mentality. Digitalists hate mysteries. Why? Because digitalists love beauty: where there is clarity, there is structure; where there is structure, there is beauty. But mystification blurs all structure into a dark indefiniteness.

For digitalists, all minds are entirely natural things that are completely open to scientific study. All minds are computations. Of course, digitalists can appeal to an enormous amount of philosophical and scientific work, across many fields, to justify the thesis that minds are computations. Unfortunately, it sometimes seems that no amount of evidence or argument for the computability of the mind will ever convince the mysterians. But there shouldn't be any need for digitalists to defend the computability of the mind right here: digitalism, like any philosophy, or like mathematics, has every right to start with its own axioms. Obviously, there are alternatives to digitalism. But those alternatives are not under discussion here – here we are studying digitalism. And you don't have to agree with the axioms of digitalism in order to study their consequences.

As your ghost replicates your brain-process, it replicates your mental life. It exactly reproduces your perceptions, your thoughts, your feelings, and your behaviors. It replicates your awareness and your self-awareness. It experiences its own self exactly as you experienced your self. It duplicates what it was like to be you.[8] For example, since your ghost brain models your body in a high level of detail, it knows what your flesh felt like when you went out running. When it replicates your run, it feels its muscles painfully contracting; it experiences the thrilling rush of endorphins. It has the same subjective awareness of running that you had when you ran. It knows what it is like to perform its actions and it can report on its inner states as you would have reported on yours.

As long as it replays your life, your fourth-generation ghost perfectly duplicates your first-person perspective. Its mental life is indistinguishable from yours. If you have a fourth-generation ghost, then somebody can set up that ghost to replay your entire life from birth to death. Your ghostly life, replayed in some computer, is mentally indistinguishable from your organic life. For all you know, your present experience is merely ghostly. Reading these very words, you might be replaying events which happened to some organic body long ago. You might *be* a digital ghost, and no amount of empirical evidence you can offer can possibly demonstrate otherwise. Anything you do to try to prove that you are not a digital ghost was something you once did before to try to make the same point. But from inside your own first-person perspective, the point is irrelevant.

6. The eternal return of the same

Fourth-generation digital ghosts are technologically possible. There are some possible future human civilizations in which any human brain-process can be perfectly recorded from birth to death, millisecond by millisecond, at the biomolecular level of detail. For the sake of argument, suppose you are living in one of those possible future civilizations. Your entire brain-process from birth to death is perfectly recorded. After you die, you remain, you persist, in your perfect neurological ghost. Since your ghost is a program, it can run after your death. It can replay your entire mental life. It can even be embedded in a loop that repeats your entire mental life over and over again for as long as possible. If your digital ghost were to be repeated, from birth to death, over and over again, then your life would be caught up in something like the *eternal return of the same.*

The ancient Greek philosopher Eudemus told his students about the eternal return like this: "If one were to believe the Pythagoreans, with the result that the same individual things will recur, then I shall be talking to you again sitting as you are now, with this pointer in my hand, and everything else will be just as it is now" (Kirk & Raven, 1957: frag. 272). The eternal return is more recently associated with Nietzsche. He uses his character Zarathustra to talk about it. Zarathustra has two animals, an eagle and a snake. They tell him that they understand his theory of eternal recurrence:

we know what you teach: that all things recur eternally and we ourselves with them, and that we have already existed an infinite number of times before and all things with us. You teach that there is a great year of becoming... this year must, like an hour-glass, turn itself over again and again, so that it may run down and run out anew. So that all these years resemble one another, in the greatest things and in the smallest, so that we ourselves resemble ourselves in each great year, in the greatest things and in the smallest. And if you should die now, O Zarathustra: behold, we know too what you would then say to yourself... "Now I die and decay" you would say, "and in an instant I shall be nothingness. Souls are as mortal as bodies. But the complex of causes in which I am entangled will recur – it will create me again! I myself am part of these causes of the eternal recurrence. I shall return, with this sun, with this earth, with this eagle, with this serpent – not to a new life or a better life or a similar life: I shall return eternally to this identical and self-same life, in the greatest things and in the smallest, to teach once more the eternal recurrence of all things."

(Nietzsche, 1885: III 13/2)

If the eternal return is true, then Zarathustra will live again after he dies. He will have a kind of life after death. His next life will physically and therefore mentally replicate his previous life. The repetition of his mental life is analogous to the repetition of your mental life by your perfect neurological ghost. When that ghost is replayed, it experiences your whole life over again. It reproduces your awareness and self-awareness. If your ghost life is replayed after your death, then that repetition reanimates your entire mental life. This at least *looks like* a kind of life after death. But if your entire ghost life were replayed, would that *really* give you life after death? To answer this question, it is necessary to answer four other questions. And all these questions will be answered.

7. Beyond digital ghosts

The first question concerns identity. Obviously, your digital ghost is *not* identical with you. Your digital ghost (when it is replayed) is merely a perfect copy of your mental life, and a copy is never identical with its original. Your present self is not identical with its corresponding momentary ghost self, and your organic life is not identical with your future ghost life. Hence the first question asks: does life after death require identity through time? Or is identity through time an illusion? Look at how Eudemus and Zarathustra phrase their statements involving identity. Rather than using the present tense, they use the future tense. Zarathustra says that the eternal cycle of causes "*will* create me again" (Nietzsche, 1885: III 13/2). He says: "I *shall* return... I *shall* return eternally to this identical and self-same life". He asserts that he *will be* identical with some future thing – not that he *is* identical with it. You cannot truly say that you *are* identical with your digital ghost. But can you truly say that you *will be* identical with it? To answer these questions, Chapter II discusses structures whose parts exist in different times – it tackles the logic of persistence.

The second question concerns the body. Your fourth-generation ghost does not have its own body. While it is a perfect *neurological* ghost, it is not a perfect *physiological* ghost. Your ghost organs are based only on your memories. Your experiences of those organs are replicated in your ghost, but the organs themselves are not. Your subjective experience of getting sick is replicated, but the activity of your immune system is not – your ghost doesn't have an immune system. Since your ghost does not really have a body, it might be argued that your ghost does not really live. It might be argued that merely mental life, the life of the brain, is not *human* life. Here is the second question: are you identical with your brain? Or are you identical with your body? To answer these questions, Chapter III carefully studies the computational anatomy of the body, the digital mechanics of the flesh.

The third question concerns autonomy. Your digital ghost is really only a *memorial*. It does not have an autonomous life of its own. Existing in a fixed historical simulation, it is trapped in your old life – it cannot actualize any new potentials. On the contrary, its very purpose is to preserve your life in a perpetually frozen form. Your digital ghost inhabits your digital grave. The third question is this: can the exact replication of your life provide you with life after death? Or does life after death require autonomy? The fourth question, which follows directly from the third, concerns solipsism. Your ghost lives in a world built only from its own memories and recorded data. Your ghost inhabits a hallucination – it lives in its own solipsistic dreamworld. Since your biography recorded the things in your external world from and only from your perspective, none of those things have their own perspectives. They do not have their own natures. Here is the fourth question: is solipsism sufficient for life after death? Or does life after death require that your self spreads out into a world filled with otherness? These questions are answered in Chapter IV. But the very process of answering them will raise disturbing questions about *our* world.

II
Persistence

8. Pipelines

You start with an empty universe, with the simplest and therefore the most boring of all possible universes. Fortunately, the next possibility is more exciting. You make a copy of your first universe, but now you add a single space-time point. Still boring, but keep going. For your next universe, you add a single feature to the point: the point can store *one bit* of information, either zero or one, off or on. Set the value of the point to zero – zero comes first. Are we having fun yet? No. Nothing is happening in these dead universes. To make something happen, you'll need to introduce some *time*.

Time starts with a point with the value zero. It changes into a second point with the value one. But why stop there? The second point changes into a third point with the value zero again. And so it goes, endlessly, with each previous point turning into some next point with the opposite value. All these points, arranged in linear order from past to future, form a *timeline*. Along this timeline, from each previous point to its next, the value gets *inverted*. Hence each previous point is linked by an *inverter* to its next point. This universe of alternating points is shown in Figure II.1. Points with value zero are white while those with value one are black. The inverters are the arrows running from point to point.

Each inverter is a trivial computer that runs a simple program: its output is the opposite of its input. This program consists of two *if-then rules:* if the previous point is off, then the next point is on; if the previous point is on, then the next point is off. Each inverter gets one bit of information from its previous point, flips it, and sends it on. Each inverter is a *pipe* through which some information flows. Specifically, it is a pipe composed of its two if-then rules. All these points, linked by pipes, form

Figure II.1 Some points in the alternating universe.

a *pipeline*. It's a channel through which data flows from the past into the future, according to a program. Of course, none of the points on this pipeline are identical – they are all distinct objects. For if they were identical, then they would have the same features – but they do not.

Figure II.1 shows a universe with one temporal dimension. As Figure II.1 shows, each pipe from one point to another is a causal connection – it is a causal arrow.[1] The input to any pipe is the cause; its output is the effect; and the if-then rules that compose the pipe are the laws that transform causes into effects. Hence the value of each previous point causes the value of the next point. Any pipeline is *causally continuous*. But the universe defined by the pipeline in Figure II.1 remains dull. If we want to make it more exciting, we need to make it more complex. One way to increase the complexity of this alternating universe is to add some *space*. So your next universe will include some space.

You can add space by replicating the pipelines in the alternating universe, like making lots of copies of a length of string in order to make a rope. But that's silly. It's far more interesting to let the spatially distinct points interact. At any moment of time, the points are lined up spatially. Each point has exactly one present neighbor above and one below. Each point and its two present neighbors pipe their values into a computer, which pipes its output to some next point. Figure II.2 shows how information flows from some points in one moment, through some computers, to some points in the next moment. The points are the squares and the computers are the arrows. Figure II.2 is fragmentary – dashed lines indicate links to other parts of the universe. No object in Figure II.2 is identical with any other – they are all distinct objects in distinct moments of time and places in space. The network in Figure II.2 displays both space and time – it is an *eternal* structure.

Any computer in Figure II.2 runs a program composed of many *if-then* rules. All these rules form a pipe, and, if you've got the pipe, then you've got the computer too. The computer *is* the pipe. Figure II.3 shows the rules that make up the program running in Figure II.2. There are eight ways that the three points on the if-side of any rule can be: 111, 110, 101, 100, 011, 010, 001, and 000. Hence any program that changes three previous point values into one next point value contains eight rules. Each rule can be pictured by means of a schematic diagram. Figure II.3 shows eight of these diagrams, which specify a complete program.[2] Each

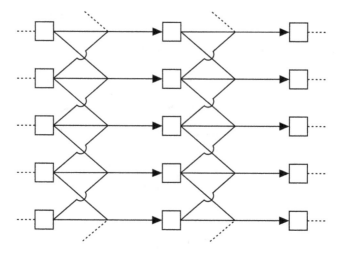

Figure II.2   Information flows through pipes.

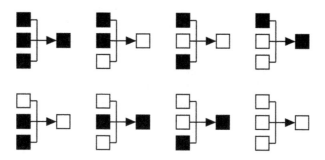

Figure II.3   The eight rules in a program.

diagram consists of the three past points on the left side (the if-side) and the one future point on the right (the then-side). The value one is black while the value zero is white. Each arrow indicates the flow of information. The second diagram on the top row in Figure II.3 shows this rule: if 110, then 0.

It's fun to design and create universes. When you add space to your universe, you specify a spatial dimension that contains some fixed number of points. You specify a temporal dimension that also contains some fixed number of moments of time. For each moment of time, you make a copy of the spatial dimension. You've now made a two-dimensional (2D) array of space-time points. No point in this array is identical with any other. Each point has its own distinct space-time location. Each point has one digital value, either zero or one. Now you select one of the three-to-one programs. You make copies of that program and you use it to pipe each triplet of previous points into its next point.

When you create this digital universe, you've made a *cellular automaton*. This cellular automaton is a big connect-the-dots structure. Of course, when you create your cellular automaton, you've also got to create its initial distribution of bit-values. You need to define its first moment of time. Figure II.4 illustrates the evolution of a cellular automaton that uses the program in Figure II.3. It starts with the small group of initially active points on the bottom (colored black) and grows temporally upwards into the future.

Each point in Figure II.4 is linked by a pipe to the point directly above it. Hence every column in Figure II.4 is a pipeline. As you go upwards in any column, you move from each point to its next point, from each point to its *successor*. From each point to its successor, data flows through a pipe. The pipe *carries* the data. But what is data? Figure II.5 shows how the same program *p* occurs between several pairs of adjacent points. From each point to its successor, this program recurs. It recurs because the pipes that carry the data also carry their own programming. After all, one of the central lessons of computer science is that programs are also data – everything is data.

A pipeline is a sequence of objects, linked by pipes, in which every pipe runs the same program. The program repeats itself from each pipe in the pipeline to the next. The program *runs through* each pipe, and

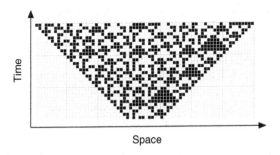

Figure II.4   A few stages of a cellular automaton.

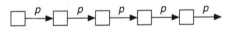

Figure II.5   The pipeline carries its own program.

it therefore runs through the pipeline. But here the sameness of the program is abstract. Just as two different sentences can express the same meaning, so two different programs can express the same functionality. The functionality of any program is its essence. Pipes preserve the functionalities of the programs that run through them – what essences they carry, they also *save*. Of course, pipelines are also pipes. They are just longer pipes made of shorter pipes stuck together.

Pipes are basic for digitalism.[3] Digitalists will define many types of pipes. For example, one very basic type includes each pipe that runs from any previous point in any digital universe to its next point. Since the next point is an updated version of its previous point, this basic type of pipe is an *updating pipe*.[4] When a pipeline is made of updating pipes, each next thing in the pipeline is just an updated version of the previous thing. It is updated by some if-then rule in the program. Another type of pipe is a *copying pipe*. All the data that goes into a copying pipe comes out the same on the other side. Copying pipes do not update any data – they preserve everything.

9. The game of life

After a while, you tire of making universes with only one spatial dimension – so you set your sights on making a digital universe with two spatial dimensions. You start with a 2D cellular automaton known as the *game of life* (Poundstone, 1985).[5] You can easily find versions of the game of life on the Internet to run on personal computers.

The basic elements of any game of life are its space-time points. At any moment, the space of any game of life is an infinite 2D grid, like an infinite chess board. The time of any game of life is a discrete series of moments or clock ticks. So the space-time of any game of life is a discrete 3D manifold (one temporal dimension plus two spatial dimensions). It is like a stack of infinite chess boards. The space-time of the game of life is often called the *life grid*. Points at distinct moments in the life grid are distinct points. And points at distinct places are distinct points. Each point in the life grid is identical with and only with itself, and it is not identical with any other point.

Each point in any game of life has eight *present neighbors* – it is surrounded by its present neighbors like the central square in a tic-tac-toe grid is surrounded by other squares. Each point has exactly one *future neighbor*. Its future neighbor is a point at the very same location in space, but shifted one moment into the future. Each point changes into or turns into its future neighbor. However, it is not identical with its future neighbor. Its future neighbor is a distinct point located at a distinct position in space-time.

Every point in the game of life is either *energized* or not. If it is energized, then its value is one; if not, then its value is zero. The distribution of energy values over all points in any life grid is its *energy field*. Over time, the energy field changes according to the game of life program, which defines the way energy flows from past points to future points. And the game of life program itself gets piped from point to point.

The game of life program consists of four if-then rules. First, if a point is energized and it has two or three energized present neighbors, then it sends one unit of energy to its future neighbor. Second, if a point is energized but it does not have exactly two or three energized present neighbors, then it sends no energy to its future neighbor. Third, if it is not energized and it has exactly three energized present neighbors, then it sends one unit of energy to its future neighbor. Fourth, if it is not energized and it does not have exactly three energized present neighbors, then it sends no energy to its future neighbor.

The if-then rules in the game of life program form an updating pipe. Points in any life grid are linked by these pipes. Each point, along with its present neighbors, is linked to its next point by one of these updating pipes. If you picture the space-time of the game of life as a stack of chess boards, rising into the future, then each column of points in that stack is a pipeline. From an eternal perspective, any game of life is just a big 3D connect-the-dots network in which the dots are points and the connections are pipes. Energy flows through these pipes. Since points in the future receive energy only from their past neighbors, the energy levels of all the points at one moment determine the energy levels of all the points at the next moment. The game of life is deterministic.

From moment to moment, the energy field changes. Vortexes and whirlpools can form in the energy as it flows through the pipes. Figure II.6 illustrates how the game of life program acts on a horizontal bar of three energized points. The horizontal bar changes into a vertical bar, which then changes back into a horizontal bar. Of course, these bars are all distinct things in distinct moments. A series of oscillating

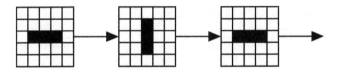

Figure II.6 Transformations of patterns on the life grid.

bars is a *blinker*. It's one of the simplest processes in the game of life. But there are other processes of much, much greater complexity. Some of those processes are universal computers (Rendell, 2002). Others can self-replicate like living cells (Poundstone, 1985: ch. 12).

10. The flight of the gliders

One of the most famous patterns in the game of life is the *glider*. The glider is something like a *wave* in the energy flow of the game of life. Figure II.7 shows five gliders, namely, Glider-1 through Glider-5. These five gliders occupy five distinct regions of space-time in their game of life. They differ from one another in their parts and in their properties; but if any things differ in their parts or in their properties, then they cannot be identical; hence no glider in Figure II.7 is identical with any other glider.[6] The gliders in Figure II.7 are all distinct gliders. And what holds for gliders holds for any things made of points in any game of life: things made of distinct points are distinct things.

Within any game of life, there is no identity through time. Since there is no identity through time in any game of life, there are no *enduring* things in any game of life. For any thing to endure, it has to remain identical with itself despite changes in its parts or properties. The concept of endurance was developed by Aristotle. But the temporality of the game of life is not Aristotelian.[7] The game of life is like the river of Heraclitus into which no one can step twice. Every object in the game of life is *ephemeral*. The game of life nicely illustrates the Buddhist principle of the *impermanence* of all things.

Affirming impermanence, various philosophers have developed a theory of persistence known as *exdurantism* (Sider, 1996, 2001; Hawley, 2001).[8] An exdurantist denies that things remain the same through time – there is no identity through time. Of course, exdurantists affirm that things *continue* through time. Continuity through time is *persistence*. Any thing that persists through time changes into, turns into, or

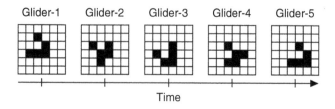

Figure II.7   The flight of the gliders.

becomes some *other* thing at some later time. It *persists into* that other thing.

Any thing that persists into some other thing is involved in some temporally extended *process*. And, just as movies are composed of photos, so processes are composed of *stages*. To say that one thing persists into another is equivalent to saying that those things are both stages of the same process. Philosophers often use the phrase *if and only if* to indicate equivalence. Thus one thing persists into another if and only if they are stages of the same process. A process is a series of stages in which each earlier stage persists into every later stage. All the stages of any process are *processmates*.

For example, all the gliders in Figure II.7 are the stages of a *glider-process*. A glider-process is a *glide*. So all the gliders in Figure II.7 are *glidemates*. Although Glider-1 is not the same glider as Glider-2, Glider-1 is in the same glider-process as Glider-2. Thus Glider-1 is a glidemate of Glider-2. Every glider is a 2D object; it is extended in space but not in time. But every glide that lasts longer than just one moment is a 3D object that is extended both in space and in time.

For digitalists, a process is a computation. More precisely, *a process is a series of stages that runs a program*. It is a timeline that runs a program. The stages are connected by pipes that carry the program from each previous stage in the process to its next stage. The program runs right through all these pipes. Hence some previous stage persists into its next stage if and only if some pipe links the previous stage to the next stage. Since a series of stages linked by pipes is a pipeline, a process is a pipeline.

11. Living in four dimensions

Some of the classical digitalists have argued that our universe is a cellular automaton.[9] This seems unlikely. Nevertheless, even if our universe isn't a cellular automaton, there are plenty of other ways for our universe to be realized by networks of computers. Our universe might be a big connect-the-dots network in which the dots are points and the connections are computational pipes. We may well be living in a digital universe (see Section 50). But that really isn't very relevant right now.

All that matters for our present purposes is that there are some very deep ways in which the physics of our universe resembles that of the game of life. One of these ways is that our universe has both spatial and temporal dimensions. Some of our best physical theories, namely, the theories of relativity, portray time as an extended dimension of change. More formally, time is a linearly ordered set of moments. These moments are linked by three temporal relations: one moment can be *earlier* than another, *simultaneous* with itself, or *later* than another. Our best physics also associates every moment of time with some volume of space. As far as digitalism is concerned, it suffices to say that space has three dimensions. Space divides into 3D regions. Our best physical theories say that our universe is at least a 4D space-time whole. Consequently, we do not live in an Aristotelian universe – Aristotelian physics is false.[10]

In our universe, any 3D material thing is a stage. Any stage fills some region of space and occupies exactly one instant of time. Any stage is extended on every spatial dimension but not on the temporal dimension. Stages are impermanent. A process is a pipeline in which each earlier stage persists into its later stages. The most basic processes are discrete. The stages of a *discrete process* form a series whose members can be counted off using the ordinal numbers (first, second, third, and so on).[11] Digitalism starts with discrete processes. Of course, mathematicians have shown how to use discrete processes to construct dense or continuous processes, and more advanced work in digital metaphysics constructs those kinds of processes; but right now, all processes are discrete.

For digitalists, all persisting things in our universe are like gliders in the game of life. Just as gliders are 3D processes made of 2D stages, so things in our universe are 4D processes made of 3D stages. If two stages of some process are located at distinct times, then they are not identical. Hence there is no identity through time in our universe. On the contrary, there is only *becoming*. And becoming is mostly self-updating. Almost every 3D material stage in our universe persists by updating itself into its next stage. It hooks up to its next stage through an updating pipe. Particles, atoms, molecules, and cells all typically persist by self-updating. Material things are stages of pipelines extended through time. Protons, planets, plants, puppies, and people are stages of pipelines.

As an illustration of becoming, consider a cell-process composed of four cell-stages. This is shown in Figure II.8. The stages do not have the same properties; hence they cannot be identical. Nor they do not have the same parts; once again, they cannot be identical. The classical digitalists confirm this lack of cellular identity through time (Moravec, 1988: 117; Tipler, 1995: 236; Kurzweil, 2005: 383). Each cell in Figure II.8 changes into the next cell at the next time; it persists into or continues into its next cell. However, continuity is not identity. No cell in Figure II.8 is identical with any other.

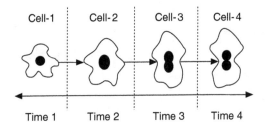

Figure II.8 Cells persist into other cells.

12. There is no identity through time

The classical digitalists all affirm the following. First, they agree with our best physical theories. They affirm the theories of relativity. They affirm that our universe contains 4D processes composed of 3D stages. Second, they affirm that human persons are human bodies (see Section 31). They affirm that human persons are neither brains, nor immaterial minds, nor mind–body composites. You are identical with your body. Third, they affirm that human bodies change their parts rapidly. Fourth, they affirm that any change in parts refutes identity through time in favor of mere continuity through time. Consequently, the theory of persistence that is most compatible with the classical digitalists is exdurantism.[12] And, given that compatibility, exdurantism is the theory of persistence that will be adopted here. Kurzweil ends his discussion of personal persistence with this rhetorical question: "So am I constantly being replaced by someone else who just seems a lot like the me of a few moments earlier?" (2005: 385).[13] Digitalists answer: *yes*.

Of course, exdurantism is controversial. But that's not much of an objection – all the theories of persistence are controversial, and we need to pick one of them. As digitalists, we certainly have the right to pick our own theory of persistence. Digital exdurantism is the doctrine that: (1) there is no identity through time; (2) temporally extended things are sequences of distinct instantaneous stages; (3) a process is a sequence of stages in which each previous stage persists into its next stage; (4) to persist is to run a program; (5) hence a process is a series of stages in which some program runs from each previous stage through some pipe to its next stage. Processes are pipelines.[14]

13. Fusion and fission

Processes are pipelines. Sometimes two distinct pipelines *fuse* together. One familiar type of fusion occurs when two cell-processes fuse into

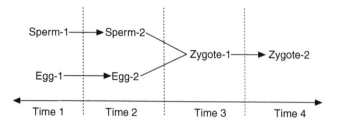

Figure II.9   Two processes fuse into one.

one cell-process. Figure II.9 shows some sperm-process and some egg-process fusing together into some zygote-process. When two processes fuse together, it usually means that their programs end with that fusion. Hence the two processes themselves end in that fusion. For example, when the sperm and egg fuse into their zygote, the sperm-program and the egg-program fuse into one new zygote-program. The sperm-process and the egg-process both end in this fusion – or, to put it bluntly, they *die*. The zygote is an entirely new thing with its own biological program. In Figure II.9, the lack of arrows into zygote-1 indicates that those links are not pipes. They blend program functionality but do not preserve it. The zygote is the first stage in some new biological process. Of course, the zygote is merely a *cell*.

Sometimes one pipeline splits into two pipelines – this is *process-fission*. Process-fission entails that persistence is not identity.[15] Fission occurs naturally when one cell divides into its two offspring cells.[16] Figure II.10 shows how zygote A fissions into two offspring B and C.

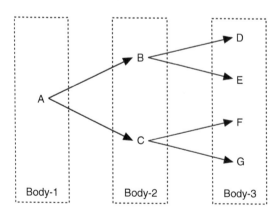

Figure II.10   A tree of dividing cell processes.

It then shows how B fissions into D and E while C fissions into F and G. Programs typically run through fission. Hence processes typically persist or continue through fission. More specifically, the program running in any cell in Figure II.10 runs through fission. The arrows from cell to cell indicate that those links are pipes. They preserve program functionality. Figure II.10 therefore contains four cell-processes. These are ABD, ABE, ACF, and ACG. These processes overlap – they all contain zygote A. And the processes ABD and ABE share their first two stages. As cell-processes divide, their continuations form an enormous branching tree. Each stage of this tree (each set of simultaneous cells) is a body.[17] The life of the body is made of the lives of its cells. Hence this tree of parallel cell-pipelines is a body-pipeline – it is a body-process. There are three bodies in Figure II.10. Each body is a distinct thing at a distinct time.

14. Bodies are stages of lives

Any process is a 4D thing composed of 3D stages. But there are many types of stages. Some but not all 4D processes are composed entirely of 3D organisms. They are composed of living things that are extended in space but not in time. For example, your body at any instant is a 3D human organism. But your body never occupies more than one moment of time. Every body is instantaneous, ephemeral, and impermanent. This is the impermanence of the flesh: from each moment to the next, your previous body ceases to exist, only to be replaced by an updated version of itself. The thing that you were at the start of this sentence is not the thing you will be at the end. From now to now, you became an other to yourself – you are a surrogate for your past bodies.

Any process runs some type of program. But there are many types of programs. Some but not all programs are *biological*. Biological programs regulate the functions of life (such as metabolism and reproduction). They are discussed in detail in Chapter III. For now, it is enough to say that biological programs are mainly encoded in the genomes of organisms. Obviously, organisms are not identical with their biological programs.[18] A process runs a biological program if and only if there is some biological program that transforms each previous stage of that process into its next stage.

When any process runs its program, that program runs through the pipes that link the earlier stages of the process to its later stages. Some but not all processes involve only updating pipes. Each previous stage of

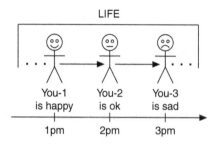

Figure II.11   Three distinct bodies in your life.

the process gets updated into the next stage. The logic of the program causes the next stage to differ from the previous stage. And some but not all processes are primarily self-contained – almost all of the action in the process occurs within its own stages. Almost all of the pipes that run into the next stage run out of the previous stage. The stages get inputs from their environments, but they don't get much in the way of external assistance. For example, ordinary bodies take in food, air, water, and heat – yet they operate primarily on their own. By contrast, a body that depends on life-support equipment for its breathing or metabolism is not self-contained.

Putting all this together, some but not all processes have these four features: (1) every stage in the process is an organism; (2) the process runs some biological program; (3) the stages persist by self-updating; and (4) the process is primarily self-contained. Any process with these features is *biologically continuous*. A *life* is a biologically continuous process; it is a biologically continuous computation; it is a series of self-updating organisms. All the bodies in one life are *lifemates*. A *human life* is a series of self-updating human bodies. Figure II.11 depicts three human bodies: You-1, You-2, and You-3. Each body is a 3D stage of a 4D body-process. This body-process is part of your life.

Perhaps it is controversial to define life in computational terms. Digitalists can, of course, defend this definition by appealing to an enormous scientific literature in fields like computational biology and artificial life. Others have argued, with great clarity, for the computability of life. Digitalists accept those arguments. But there shouldn't be any need for digitalists to defend those arguments here: for digitalists, the decision to define life in computational terms is axiomatic. Since we are presently studying digitalism, we will assume the digitalist axioms, and work out their conclusions.

15. Your counterparts in other times

Your life is a series of self-updating bodies. Your life is a process, and all the bodies in your life are your lifemates. At any moment of your life, you are your own present lifemate. As Figure II.12 shows, you *are* you. You are identical with your present lifemate. You are the same body and person as your present lifemate. But that's the beginning and end of your personal identity – you aren't identical with any other bodies in your life. If you are not the first body in your life, then you have some past lifemates. Suppose *Younger You* is one of them. As Figure II.12 shows, you *were* Younger You. But you are neither the same body nor the same person as Younger You. On the contrary, you and Younger You are two distinct bodies – two distinct people. If you are not the last body in your life, then you have some future lifemates. Suppose *Older You* is one of them. As Figure II.12 shows, you *will be* Older You; but you are neither the same body nor the same person as Older You. On the contrary, you and Older You are two distinct bodies – two distinct people.

All processmates are *temporal counterparts*.[19] More specifically, all life-mates are temporal counterparts. Temporal counterpart theory shows how to define the conditions in which tensed statements are true. It defines the *truth-conditions* of tensed statements. Most declarative statements are tensed. More precisely, any statement with a subject-predicate form is tensed. Here are three distinct subject-predicate state-ments: "I am in the woods", "I was in the woods", and "I will be in the woods". Since they are tensed, their truth-values are defined using tem-poral counterpart theory. Temporal counterpart theory is illustrated by the bodies in Figures II.11 and II.12. Tense by tense:

The Past. Any statement a body makes about itself in the past tense is made true or false by its past counterparts. For any predicate P, any

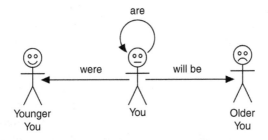

Figure II.12   Some temporal relations among your bodies.

statement of the form "I was P" is true when said by a body now if and only if one of its past counterparts *is* P. For example, if P is the predicate "happy", then the past tense statement "I was happy" is true when said by you at 2pm if and only if you have some past counterpart who *is* happy in *its* earlier time. And since You-1 in Figure II.11 is happy, the statement is true.

The Present. Any statement a body makes about itself in the present tense is made true or false by its present counterpart. For the present tense, temporal counterpart theory is trivial – every body is its own present counterpart. Hence for any predicate P, any statement of the form "I am P" is true when said by a body now if and only if that body itself is P. For example, "I am ok" is true when said by you at 2pm if and only if you at 2pm are ok. And, since You-2 at 2pm in Figure II.11 is ok, that statement is true.

The Future. Any statement a body makes about itself in the future tense is made true or false by its future counterparts. For any predicate P, any statement of the form "I will P" is true when said by a body now if and only if one of its future counterparts is P. For example, the future tense statement "I will be sad" is true when said by you at 2pm if and only if you have some future counterpart who *is* sad in *its* later time. Alas, since You-3 in Figure II.11 is sad, the statement is true.

16. Attachment breeds suffering

A life is a biologically continuous series of bodies. Within any life, each previous stage *lives into* the next stage. Sadly, every life eventually ends in death. Since death is the *end* of life, it is impossible for any organism to *live through* death. It is therefore impossible for any organism to live into anything that comes after its death. Does this mean that all is lost? No, it does not. You might have future counterparts in lives that exist after your present life has ended. If you persist into one of your own post-mortem bodies, then you will *live again* after your death. You will *live after* your death. Digitalists will argue that you will live again after your earthly death. It's your only hope.

One illustration of life after death is provided by the eternal return of the same (Section 6). For digitalists, the eternal return is merely a thought-experiment – you don't need to believe in the eternal return to see how it illustrates life after death. According to the eternal return, nature has a cyclical form: the entire physical structure of our universe gets repeated, over and over again, endlessly, into both the past and future. On this view, the physical structure of our universe is one small part of nature. Nature is just one big program loop: (1) run our universe from start to finish; (2) go back to step one. Since it runs a program, nature is a process. Nature is one big 4D process composed of 3D stages.

Since your life is a part of our universe, the entire structure of your life gets repeated, over and over again, endlessly, into both the past and future. Your *eternal career* is the entire series of your repeated lives. It is just one big program loop: (1) run your earthly life from start to finish; (2) go back to step one. Since your eternal career runs this program, your eternal career is a process. Recurrence computationally stitches all your lives together, even though they are separated by huge temporal gaps (Section 97). Your 4D eternal career includes infinitely many exact copies of your 4D life. Since each of these 4D lives divides into 3D bodies, your eternal career is a temporally ordered series of 3D bodies. Some of these bodies are in the same life; but most are in different lives. Nevertheless, within your eternal career, every body is a temporal counterpart of every other body.

More precisely, every body in your eternal career is a *recurrence counterpart* of every other body. You are identical with exactly one of your recurrence counterparts, namely, your present body. Of course, you *are not identical with* any other recurrence counterparts – they are all distinct bodies in distinct times. But you *were identical with* each of your past recurrence counterparts, and you *will be identical with* each of your future recurrence counterparts. You can truly say: I will be born again, I will live again, and I will read this sentence again. As far as identity goes, recurrence is just like ordinary earthly life: just as you *are not identical with* any past or future bodies in your earthly life, so also you *are not identical with* any past or future bodies in your eternal career. Identity is a purely logical relation. As such it is so thin, so devoid of content and meaning, that it plays no positive role in any theory of either ordinary life or life after death. Identity is irrelevant.

Any difference between your earthly life and your eternal career involves, not identity, but *continuity*. Although your earthly life is biologically continuous, your eternal career is not biologically continuous – each life in your eternal career is separated by death. You can truly say: as I have died infinitely many times before, so I will die infinitely many times again. You do not *live into* any of your post-mortem recurrence counterparts. But you do *remain into* your post-mortem recurrence counterparts (they all carry perfect information about you). You persist into them. The programmed repetition of your life, within the programmed repetition of the universe, ensures that your eternal career has a kind of computational continuity. And, just as your biological continuity spans sleep, so also that computational continuity spans death. Digitalism will not argue for the eternal return; but it will argue for many types of computational continuity that span death.

Of course, you might still object that your *desire* for life after death cannot be satisfied by the existence of any post-mortem temporal counterparts. But if your desire for life after death cannot be satisfied by those counterparts, then your desire for life *after* death conceals a hidden desire for life *through* death. Unfortunately, it is logically impossible for you to live through your death. If you want to live through your death, then your desire is self-contradictory. You're bound to be perpetually frustrated. Your desire for life after death ought to be satisfied by your post-mortem temporal counterparts. Specifically, it ought to be satisfied by the fact that you *will be* identical with every one of them.

On the one hand, if you want to live *through* your death, then your desire involves an *attachment to presence*. This attachment to presence generates the illusion of identity through time. Since the present perpetually flows away from itself in the river of becoming, your attachment to presence, your belief in identity through time, can only breed *suffering*. On the other hand, if you want to *overcome* your death, if you want to *prevail over* your death, if you want to *surpass* it, then you have to surrender your attachment to presence. You have to surrender your painful desire to stay the same, to remain self-identical. Perhaps this point can be affectively reinforced by the Buddhist verse that ends Parfit's *Reasons and Persons* (1985: 503). The verse runs like this:

> The mental and the material are really here,
> But here there is no human being to be found.
> For it is void and merely fashioned like a doll,
> Just suffering piled up like grass and sticks.[20]

17. Your ghostly counterparts

Your brain-process is a 4D sequence of 3D brains. Digitalists argued in Chapter I for the technological possibility of fourth-generation digital ghosts, which exactly replicate your entire brain-process, from birth to death. If your fourth-generation ghost is replayed, then your brain-process is reanimated, so that your mind-process is reanimated. Your ghost will replay your entire mental life. Since your ghost carries information about the life of your brain, which is a part of your life, your life partly remains in your ghost.

Your ghost brains are temporal counterparts of your earthly brains. If your brain-process gets replayed by your ghost, then all those simulated brains are *ghostly counterparts* of your earthly brains. And, since the simulation runs in the future, they are all future counterparts. Each one of your earthly brains will be identical with each of its future ghostly counterparts. You can truly say that a *part* of you (your brain right now) will be identical with your corresponding ghost. But will *you* be identical with your ghost? Will you persist into your ghost? Can your digital ghost provide you with life after death? If you are your brain, then the answers are all *yes;* otherwise, they are all *no.* Digitalists clearly need to answer these questions. Consequently, for digitalists, the next task is to study the logic of the body – what is the structure of the flesh?

III
Anatomy

18. Our finite earthly cells

Any earthly cell has some finite size – it occupies some finite spatial volume.[1] Since atoms have only finite sizes, every cell contains only finitely many atoms. And since any atom has only finitely many ultimate particles, every cell contains only finitely many ultimate material parts. So the number of material parts of any cell is finite.

It may be that our space is infinitely divisible; nevertheless, finitely many atoms cannot be arranged in infinitely many ways in any finitely sized cell. The atoms in earthly cells are compressed together. Any atom can form only finitely many different types of covalent bonds with other atoms; hence there are only finitely many ways that the atoms in any cell can be covalently combined. And those covalent bonds determine the shapes of the molecules that bond via the van der Waals force. The number of possible spatial relations among the atoms (and molecules) in any earthly cell is finite.

It may be that our time is infinitely divisible; nevertheless, infinitely many chemical reactions cannot take place in any finite period of time. Any finite period of cellular life is ultimately divisible into only finitely many smallest intervals. The changes of the cell are neither continuous nor dense – they are discrete. Change within the cell runs at a finite clock speed. Since the cell at any time has only finitely many possible arrangements of parts, it follows that, from one clock tick to the next, the cell can change in only finitely many ways. For the complete description of the changes of any cell, it is not necessary to use differential equations; on the contrary, difference equations are sufficient.[2]

All the numbers that characterize the complexity of any cell are finite. No real numbers or even rational numbers are required for the complete

30

description of any cell. Any cell can be perfectly described by finitely many natural numbers (the whole numbers, starting from zero). But any natural number is equivalent to finitely many binary digits (bits). Thus every cell can be perfectly described by finitely many bits. Sagan (1995: 987) says any *human* cell encodes about 10^{12} bits. Thus any human cell encodes about 1 terabyte of information. This is a large but finite number. Today (circa 2013), you can buy hard drives that store far more than that. Any cell is only finitely complex. It is a discrete dynamical system with finitely many trajectories in a finite phase space.

19. Cells have minds

Any cell has many layers of organization. These can be roughly characterized as the physical, the chemical, and the biological. The physical level involves particles and atoms while the chemical level involves molecules. But what about the biological level? Systems biologists and computational cell biologists picture the cell as a network of interacting nano-machines.[3] These nano-machines are often referred to as *nanites*. At the biological level, the cell is a society of interacting nanites. The cell is a connect-the-dots network in which the dots are nanites and the connections are biological interactions.

A nanite is a machine that is realized or implemented by a system of chemical or physical objects (typically DNA, RNA (ribonucleic acid), and proteins).[4] Realization usually conceals enormous internal detail at some lower functional level. Thus a nanite with only a few biological states may be realized by a molecular assembly with enormously many chemical or physical states. The biological states are classes of biologically equivalent chemical or physical states. For example, an ion channel has two biological states: open or closed; but each biological state is a class that includes many chemical or physical states. Likewise a receptor is either active or inactive; a gene is either being transcribed or not. Higher-level nanites are typically realized by networks of lower-level nanites.[5]

It is customary to divide the cellular nanite network into three relatively isolated subnets: the *signaling, regulatory,* and *metabolic* subnets (one might add a *cytoskeletal* subnet). Of course, this division is somewhat vague – these networks overlap and are only roughly defined. Nevertheless, this tripartite scheme is often used in cell biology. For digitalists, the signaling and regulatory networks are especially interesting. These networks recognize patterns and make informed decisions (Steinhart, 2001: sec. 3.5).[6] These networks are the *mind of the cell.*

As you'd expect, cells have very small minds. Cellular minds are micro-minds with very little intelligence. Still, intelligence is something that cells clearly do have – they survive on their own in the wild, they live by their wits.

20. The form of the cell

Although viewing the cell as a network of nanites already involves considerable abstraction, it will be useful to move to an even higher level of abstraction. And, to introduce that higher level, it will be helpful to consider a trivial cartoon example – a toy cell. This toy cell is a trivial swimming cell. Its life just consists of swimming around.

The life of the cell involves some *inputs*, *states*, and *outputs*. At any time, this trivial swimming cell gets some inputs from its environment. Any input to the cell is realized by some society of nanites moving into the cell. At any moment, the cell is in some internal state, which is realized by the society of nanites inside the cell. And the cell always produces some output, which is some society of nanites it sends to its environment.

Our little swimming cell lives in a very simple liquid environment. Sometimes this liquid is light and sometimes it is dark. However, sometimes this liquid becomes acidic. So the set of possible inputs to the cell is {Light, Dark, Acid}. The cell has a start state at the beginning of its life and an end state when it dies. While it is alive, it can only be happy or sad. So its set of possible states is {Start, Happy, Sad, End}. Its outputs are the directions in which it swims. It can swim straight, turn right, or turn left. When it dies, it just stops moving. So its outputs are {Straight, Right, Left, Stop}.

From one moment to the next, the cell changes. The ways it changes are defined by its if-then rules. Each rule links the current state and input of the cell to its next state and output. Each rule has this form: if the cell gets an input while it is in some current state, then it changes to its next state and it produces an output. Each rule is realized by the motions of the input and state nanites. Each rule specifies a way that those motions cause the cell to arrange its nanites into its next state and to produce some nanites as output.

Here are the rules: (1) if the cell gets Light while it is Starting, then it becomes Happy and swims Straight; (2) if it gets Darkness while it's Starting, then it becomes Sad and swims Straight; (3) if it gets Light while Happy, then it stays Happy and it swims Straight; (4) if it gets Light while Sad, then it turns Happy and it swims Left; (5) if it gets

Darkness while Happy, then it turns Sad and swims Right; (6) if it gets Darkness while Sad, then it stays Sad and swims Straight; (7) if it gets Acid while Happy, then it Ends and Stops swimming; (8) if it gets Acid while Sad, then it Ends and Stops swimming.

Figure III.1 displays these rules as a *state-transition diagram*. Each circle indicates a state. Each labeled arrow from a circle to a circle depicts a rule. Just as there are eight rules, so there are eight arrows. Each arrow is labeled with an input/output pair. For instance, Figure III.1 contains an arrow from the circle labeled Start to the circle labeled Happy. The arrow is labeled with light/straight. This labeled arrow represents the first rule: if the cell gets Light while in its Start state, then it becomes Happy and swims Straight.

The if-then form of each rule suggests that these rules encode two associations between the items in the if-part and the items in the then-part. The first association links every (input, state) pair with some next state. Mathematically speaking, this association is a *function* that maps each (input, state) pair onto some next state. Functions specify unambiguous links from things in one set to things in some other set. For instance, in a monogamous culture, the marriage function unambiguously associates each man in the set of husbands with exactly one woman in the set of wives. For cells, the first association is the *transition function*. The second association is a function that maps each

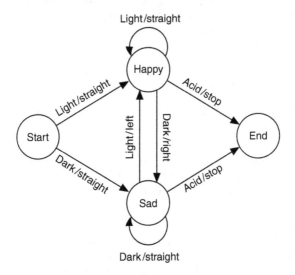

Figure III.1 The state-transition diagram for the swimming cell.

(input, state) pair onto some output. This second association is the *production function*. Since these associations are both functions, cells are *deterministic*.[7] However, this determinism does not preclude freedom. For digitalists, determinism is *compatible* with freedom.

According to *compatibilism*, an agent is free when, and only when, it acts out of its own deterministic nature. Freedom is *self-determination*. Cells are agents, and the nature of any cell is defined by its if-then rules. When the changes in any cell are governed by its own if-then rules, the cell is acting freely. However, when an agent is being *coerced* by some alien power, when it is being compelled to act against its own nature by some external force, then it is not free. Cells can be coerced in many ways. They can be pushed around by mechanical forces, poisoned or altered by chemicals, by heat, by radiation. Or they can be infected by viruses. When a virus hijacks a cell, then the if-then rules encoded in the virus begin to govern the cell. The alien nature of the virus coerces the cell to do things that it would not do on its own. Unfortunately, that sick cell is no longer free.

All the elements that are needed to formally define cells are now in place. These elements are the set of inputs; the set of states; the set of outputs; the transition function; and the production function. Since cells are finite, these sets are all finite, and each function contains only finitely many links. More technically, these five elements form the quintuple (input set I, state set S, output set O, transition function F, production function G). All the if-then rules realized by the cell are mathematically compressed into this quintuple. Since it incorporates all those rules, it is the *nature* of the cell.

21. The life of the cell

Although real earthly cells are far more complicated than the trivial swimming cell, they are nevertheless only finitely complex. So this abstract analysis applies equally well to real earthly cells (and, even with infinitely complex cells, this analysis still works). This analysis is precise, despite its high level of abstraction. Digitalists conclude that every earthly cell realizes or implements a cellular nature of the form (I, S, O, F, G).

Every earthly cell has a nature. Its nature includes all possible cellular inputs, states, outputs, transitions, and productions. And its nature remains invariant through all the natural changes of the cell. It remains the same as long as the cell is free. From the point of view of computer science, the nature of the cell is the *cell-program*. From the point of view of philosophy, this nature is the *form* of the cell. Digitalism brings these two perspectives together: *the form of the cell is the cell-program*. Every cell *implements* or *realizes* its program. And since any physical system that implements a program is a *machine*, every cell is a machine. Technically, since the number of states in any earthly cell-program is finite, it follows that *every earthly cell is a finite state machine*.

A cell is a temporally unextended 3D thing. A series of cells is any temporally ordered set of cells. Every series of cells is 4D. A series of cells *runs a program* if and only if every previous cell in that series is piped into its next cell. The cell-program flows from each previous cell through some pipe to the next cell. The pipes involved in the lives of cells are updating pipes (see Section 11). These pipes are composed of the if-then rules in the cell-program. Since updating pipes link 3D cells into biologically continuous 4D cellular lives, it will be helpful to look at them in more detail. Updating pipes define *successors*. If an updating pipe runs from some previous cell to some next cell, then that next cell is a successor of the previous cell.

The successor relation among cells is defined in terms of the successor relation among the *configurations* of cells. A later cell is a successor of an earlier cell if and only if the configuration of the later cell is a successor of the configuration of the earlier cell. The configuration of any cell includes exactly its program, input, state, and output. Formally, the configuration of any cell is the quadruple (program p, input i, state s, output o). Any later configuration is a successor of an earlier configuration if and only if (1) the later program is identical with the earlier program; (2) the later input is any possible input from the environment; and (3) the later state and output are derived from the earlier state and input by some rule in the earlier program.[8] Since the environment can usually supply many possible inputs, any earlier cell can usually have many possible later successors.

At the cellular level, a *life* is any biologically continuous series of cells (see Section 14). The life of any cell is a process, and all the cells in that process are processmates – they are temporal counterparts. Each cell *was* the earlier cells in that life and it *will be* the later cells in that life. Table III.1 illustrates a biologically continuous series of cells running the program from Figure III.1. Time runs from left to right. Thus Cell-4 is a successor of Cell-3, which is a successor of Cell-2, which is a successor of Cell-1. All these cells form a pipeline. They are 3D stages of one biologically continuous 4D cellular life. Each earlier cell is linked to its successor by an if-then rule in the cell-program p. Rule two changes Cell-1 into Cell-2; rule four changes Cell-2 into Cell-3; and rule seven

Table III.1   A series of cells running the program from Figure III.1

	Cell-1	Cell-2	Cell-3	Cell-4
Program	p	p	p	p
Input	Dark	Light	Acid	None
State	Start	Sad	Happy	End
Output	None	Straight	Left	Stop

changes Cell-3 into Cell-4. These four cells are processmates. They are all temporal counterparts: each cell is itself; but Cell-1 will be Cell-4; and Cell-4 was Cell-1.

When one cell changes into another cell, its matter changes. And while the matter of the cell changes over time, the form of the cell remains the same. It is a constant pattern, an exactly conserved dynamical nature. The persistence of the life of any cell requires the persistence of its form. When the form of the cell is altered, the life of the cell ends. Since the form of the cell is invariant through all natural changes, it seems reasonable to say that the form of the cell is associated with some part of the cell whose material structure remains invariant through all natural changes. That part is the *genome* of the cell. For eukaryotic cells (like ours), the genome is found in the nucleus and mitochondria.[9] Of course, since the genome is a material thing, the form of the cell is not identical with the genome. On the contrary, the form of the cell is *implicit in* or *encoded in* the genome.

Since the genetic code is ambiguous, many materially distinct genomes can operate in the same way; two materially distinct genomes can be functionally equivalent. Hence there is a *functional equivalence* relation among genomes. A *genotype* is a collection of functionally identical genomes. Two cells share the same form if and only if they share the same genotype. Of course, genotypes are not identical with cell-forms. On the contrary, cell-forms are far more abstract. Genotypes merely encode cell-forms. And they encode them without ambiguity: one genotype encodes one and only one cell-form.

22. Digital cells

Since any cell is only finitely complex, it can be *exactly simulated* by some finite digital computer. Since digital computers can be made out of many kinds of stuff, cells can be made out of many kinds of stuff. All cells are *multiply realizable* – they are *substrate independent*. A *bion* is an artificial computer dedicated to cell simulation. Bions might be realized by semiconductors like silicon or gallium arsenide; they might be realized by organic compounds like carbon nanotubes or sheets of graphene. Bions may employ specialized hardware, such as the *cytomorphic electronics* of Sharpeshkar (2010: chs. 22–4). But specialized hardware is not functionally required. Since any finite program can be run by some classical von Neumann machine, bions are just such machines. Bions can be realized by bigger versions of the computers we use on our desks.

The finite complexity of cells ensures that it is technically possible to make a bion that exactly replicates any cell. For every configuration of any cell, it is technically possible to set up a bion so that it has that very same configuration. Thus set up, the bion is *functionally isomorphic* to the original cell – it is a perfect copy of the cell. Consequently, the configuration of the bion *is identical with* the configuration of the cell. Since cells are finitely computable, researchers have been building bions that ever more closely approximate cells. These include E-CELL and Virtual Cell.[10]

Any human cell has zillions of possible inputs, states, and outputs. And it has zillions of if-then rules in its program. But moving to bigger numbers doesn't change the form of the cell. The concept of a cell-program as a nature (I, S, O, F, G) can be extended to any desired degree of complexity. Every human cell runs a cell-program. And every human cell has a configuration (p, i, s, o). The configuration of any human cell is exactly encoded by some finite bit string. Every possible life of every possible human cell can be exactly simulated by some process running on some finite digital computer.

Digitalists insist on this exactitude. This means that digitalists are *mechanists* about cells. For digitalism, it is axiomatic that cells are machines. Nevertheless, digitalists need not insist that we will ever achieve this exactitude in our simulations of cells and their lives. On the contrary, our human efforts to model cells and their lives may never get beyond the merely approximate. Perhaps the exact computational models of cells and their lives will remain forever inaccessible in some Platonic world of mathematical purity. But so what? For the digitalist, the forms in that Platonic world provide ideals to strive for as they try to make better and better artificial models of cells and their lives.

23. The brain computes

The brain is a network of interconnected nerve cells (*neurons*). Of course, it also includes other kinds of cells (like glial cells), but most research so far has focused on neurons. Neurons are linked to other neurons by *synapses*. Hence the brain is a connect-the-dots network in which the dots are neurons and the connections are synapses. Here is an argument that the brain is a computer: each cell in the brain is a computer; the brain is a network of cells; hence the brain is a network of cellular computers; but any network of computers is also a computer; therefore, the brain is a computer.

Any computer has some processing power and some storage capacity (its memory). The processing power of any computer is stated in operations per second while its storage capacity is just stated as a number of bits. The classical digitalists all give estimates of the processing power and storage capacity of the brain (Moravec, 1988: 59–61; Tipler, 1995: 22–3; Kurzweil, 2005: 123–7). Although these estimates are similar, they are not the same. The largest numbers are used here. The brain contains 10^{11} neurons. Each neuron has up to 10^5 synaptic connections to other neurons. Hence the brain contains 10^{16} synapses. Each synapse stores up to 10^4 bits. Hence the storage capacity of the brain is at most 10^{20} bits. Each neuron performs 10^6 operations per second on its internal data. Hence the brain as a whole performs 10^{17} operations per second.

Since all the numbers describing the brain are finite, the brain is only finitely complex. It has a finite set of possible inputs. These are the signals entering it through its sensory nerves (the nerves coming from the sensory and other organs). The brain has a finite set of possible states. These are its internal connect-the-dots networks of cells. The brain has a finite set of possible outputs. These are the signals leaving it through its motor nerves (the nerves that go out to the muscles and other organs). The transition function of the brain maps all its possible (input, state) pairs onto its states. The production function of the brain maps all its possible (input, state) pairs onto its outputs.

All these sets define the program being run by the brain. This brain-program has the form (I, S, O, F, G), where the sets and functions are all specific to the brain. This program is the *form of the brain*. Since these sets are all finite, the brain-program is only finitely complex. Hence the brain is a finite state machine. Since any brain is only finitely complex, it can be *exactly simulated* by some digital computer.[11] And since every digital computer is multiply realizable, the brain is multiply realizable. Brains are substrate independent. Any organic brain can be perfectly replicated in silicon.

The brain is a network of cellular computers; but every cellular computer is a network of molecular machines (it is a network of nanites). Every cell in the brain performs intelligent computations – every cell in the brain has a mind. The *mind of the brain* is the sum of the minds of its cells. Hence the mind of the brain is a social network of nanites. The mind of the brain is identical with this network of nanites. Since this nanite network does not include all the molecules in the brain, it is smaller than the whole brain. It is a proper part of the brain and is not identical with the brain itself. But the mind of the brain is not the only mental part of the body. The brain is not the only intelligent part of the flesh. The mind of the body is bigger than the mind of the brain. The brain is not the omniscient and omnipotent ruler of the body. The body is not a monarchy.

24. The nervous system computes

The *nervous system* consists of the central nervous system (CNS) plus the *peripheral nervous system* (PNS). The CNS includes the brain plus the spinal cord and the cranial nerves. The PNS divides into the somatic and autonomic systems. The autonomic system divides into the sympathetic system, parasympathetic system, and enteric nervous system (ENS). The sense organs can also be included in the PNS. Probably the endocrine system should be added to the PNS as well.

One of the most interesting parts of the PNS is the ENS (Gershon, 1998). The ENS is a complex neural network wrapped around the digestive organs. The ENS contains over 100 million neurons organized into layered structures similar to those found in the brain. An overview of cognition in the ENS is given here. For a more detailed and technical account, see Steinhart (2001: sec. 3.2). The ENS perceives the contents of the gut via sensory channels which resemble taste, smell, and touch. It contains many neural mechanisms which are thought to be involved in learning and memory. It encodes an intelligent system of dispositions that associate perceived inputs with actions. On the one hand, the brain often controls parts of the ENS; on the other hand, the ENS often controls parts of the brain. The ENS is not the slave of the brain. On the contrary, the ENS has a mind of its own. The mind of the ENS is the sum of the minds of all the cells in the ENS.

The ENS is a part of the PNS; but the PNS and the CNS are parts of the nervous system. The nervous system is a finitely complex cellular network. It has finite sets of possible inputs, states, and outputs; it has a finite transition function and a finite production function. These sets make up the neural-program. It has the form (I, S, O, F, G), where the sets and functions are all specific to the nervous system as a whole. This neural-program is the *form of the nervous system*. Since all the sets and functions in the neural-program are finite, the nervous system is a finite state machine. Consequently, it can be exactly simulated by a digital computer. Nervous systems are substrate independent. Any organic nervous system can be perfectly functionally duplicated in silicon.

Every cell in the nervous system has a tiny cellular mind. The *mind of the nervous system* is the sum of the minds of its cells. The mind of the nervous system includes the minds of the CNS and PNS. Since the nervous system is bigger than the brain, its mind is bigger than the mind of the brain. The mind of the nervous system is a network of molecular nano-machines distributed throughout every nerve cell in the body. The mind of the nervous system *permeates* the flesh. It saturates your body from your fingers to your toes. But the nervous system is not the only intelligent part of your body. Your mental life is richer than the mental life of your nervous system. The mind of the body is bigger than the mind of the nervous system.

25. The immune system computes

Your *immune system* defends your body against infection. It contains about 10^{12} cells, distributed in diffuse networks throughout the flesh. The defensive work of the immune system involves an enormous amount of intelligence. An overview of this immunological intelligence is given here. For a detailed and technical account, see Steinhart (2001: sec. 3.4). The immune system contains sensory networks which perceive the entry of infectious agents (*antigens*) into the body. These sensory networks perform complex pattern-recognition and classification tasks. Once an antigen has been classified, these sensory networks arouse other parts of the immune system to respond.

Often the immune system encounters novel antigens. When that happens, the *adaptive immune system* may be aroused. B-cells in the adaptive immune system evolve antibodies to defeat the antigen. These B-cells run an evolutionary algorithm in which they modify their own genes. This evolutionary algorithm is a *learning* algorithm. By running it, the B-cells evolve genes which encode the most effective antibodies for fighting the antigen. They recursively improve their own genes. Once these genes are evolved, they are stored in long-lasting B-cells. The library of evolved genes for antibodies is the *memory* of the immune system (Benjamini et al., 1996: ch. 6). Immunological memory is the basis for *acquired immunity* (which includes the immunity gained from vaccination). When the immune system encounters an antigen, it scans its memory for the most effective response. If it has seen that antigen before, it reactivates the B-cells which encode the most effective antibodies against it. Those reactivated B-cells multiply and quickly overwhelm the antigens. After one infection, your immune system learns how to defeat the antigen.

Your immune system makes intelligent decisions based on its experience. On the basis of its pattern-recognition mechanisms and learning algorithms, it uses its old memories to defeat entirely new threats. This is the basis for vaccination. For example, once vaccinated with the cowpox virus, your immune system learns how to overwhelm the smallpox virus. More technically, your immune memory, like your neural memory, is associative and content-addressable. The immune system can be *trained* to respond to stimuli using Pavlovian classical conditioning. Immunological memories have well-defined semantics.[12] Your immune system can make mistakes, which manifest themselves in allergies and auto-immune disorders. Your immune system perceives, recognizes, learns, remembers, deliberates, decides, and acts. It has beliefs and desires.

Every cell in the immune system has its own little mind. The *mind of the immune system* is the sum of the minds of its cells. Although some aspects of the immune system are partly controlled by the brain, most of the activity of the immune system is entirely independent of the brain. The immune system can control the brain. Your immune system has a mind of its own, which often dominates the mind of the brain.

Your immune and nervous systems are parts of your neuro-immune system. Your neuro-immune system is a finitely complex cellular network, which runs a finitely complex neuro-immune-program. It has the form (I, S, O, F, G), where the sets and functions are all specific to the neuro-immune system. This program is the *form of your neuro-immune system*. Since all the sets in that program are finite, your neuro-immune system is a finite state machine. It can be exactly simulated by a digital computer. Any organic neuro-immune system can be perfectly functionally duplicated in silicon.

Every cell in the neuro-immune system has a molecular mind. The *mind of the neuro-immune system* is the sum of the minds of those cells. It is a network of molecular nano-machines distributed throughout the body. The mind of the neuro-immune system includes the mind of the nervous system and the mind of the immune system. Hence the mind of the neuro-immune system is bigger and more intelligent than each of its component minds. But the neuro-immune system is not the only intelligent part of the body. The mind of the body is bigger than the mind of the neuro-immune system.

26. The body computes

Every cell in the body performs intelligent computations at the molecular level – every cell in the body contains a mind. The mind of the
cell is the intelligent nanite network inside of the cell. The *mind of the
body* is the sum of the minds of all the cells in the body. The mind of
the body contains the neuro-immune mind. However, since the body
contains cells that are not in the neuro-immune system, the mind of
the body is larger than the neuro-immune mind. Simply put, the *mind*
is the mind of the body. Since the mind is a network of cells, it runs a
mind-program. It has the form (I, S, O, F, G), where the sets and functions are defined by integrating the programs of all the cellular minds
in the body. This mind-program is the *form of the mind*. It is an abstract
mathematical structure.

The mind is the part of the body that computes. It is an *organ* of
the body (Cuthbertson et al., 1996). The mind is ultimately a material connect-the-dots network in which the dots are molecular nano-
machines and the connections are their functional interactions. The
mind does not emerge from or supervene on that nanite network – it
is *identical* with that nanite network. The mind is not the software of
either the body or any proper part of the body. The mind is the visceral
hardware that runs the mind-program.

While the mind is concentrated in the neuro-immune system, it is
also distributed throughout the rest of the flesh. The mind penetrates
every cell in the body; the mind is a subtle network that permeates and
animates the flesh; the flesh is *saturated* with mentality. Thinking occurs
in every cell in the body at the molecular level; the flesh boils with
carnal thought. However, the mind is not the entirety of the body. Many
parts of the body are not parts of the mind (for example, the lipids in
cell walls; the water in cells; the cellular molecular networks involved in
only metabolism or shape).

Since the mind is a physical part of the body, it is housed inside of
the body. The mind is *embodied*. However, since the mind is woven
into every cell at the molecular level, it is not possible to separate the
mind from the body. It is not possible to surgically extract the mind
from one body and transplant it into another body. Starting with Locke
(1690: II.27.15), many philosophers have imagined cases in which bodies exchange minds. Some philosophers have said that swapping brains
is swapping minds (Puccetti, 1969); however, since minds are not brains,
brain swapping is not mind swapping (Steinhart, 2001). Minds cannot
be moved from body to body; bodies cannot swap minds.

However, since copies can be made of cells, copies can be made of minds. And those copies need not be made out of the same stuff as their originals: networks of nanites are substrate independent; but the mind is a network of nanites; hence the mind is substrate independent. Earthly minds can be exactly replicated by finite digital machines.

27. Our finite earthly bodies

Since cells are machines, the body is a machine. It may be helpful to give an explicit argument for this thesis. The *Network Argument* goes like this: (1) Since every body is a finite network of cells, and since cells are finitely complex machines, every body is a finite network of finitely complex machines. (2) And since cells interact in only finitely complex ways, every body is a finitely complex network of finitely complex machines.[13] (3) But a finitely complex network of finitely complex machines is itself a finitely complex machine. Consequently, (4) every body is a finitely complex machine.

Any body can be exactly described by finitely many bits of information. It is functionally equivalent to a digital computer with finite memory running a finite body-program.[14] And so every life of every body has an exact digital simulation. Just as they are mechanists about cells and their networks, so digitalists are mechanists about bodies: bodies are machines. For digitalists, this position is axiomatic. Perhaps objections can be raised against the mechanist conception of bodies. Nevertheless, since we are studying digitalism, we will assume the digitalist axioms, and reason from them.

28. The form of the body

All the mechanical features of the body are defined in terms of the mechanical features of its cells. Every body has some *body-input*. Its body-input is a vector composed of all the inputs from its environment to its cells. Every body has some *body-output*. Its body-output is a vector composed of all the outputs from its cells to its environment. Every body has some *body-state*. Its body-state includes the states of all the cells in the body as well as all the inputs and outputs exchanged among them. Every body runs a *body-program*. Its body-program is the network integration of its cell-programs.

Since all programs have the form (I, S, O, F, G), the body-program has that form. More precisely, any body-program is a quintuple (I_B, S_B, O_B, F_B, G_B). I_B is the set of all possible body-inputs. S_B is the set of all possible body-states. O_B is the set of all possible body-outputs. F_B is a function that associates every pair (body-input, body-state) with some new body-state. G_B is a function that associates every pair (body-input, body-state) with some body-output. Since F_B and G_B are both functions, bodies are *deterministic*.[15] However, this determinism does not preclude freedom. For digitalists, determinism is *compatible* with freedom.

All the sets in the body-program are finite. Since they are all finite, the body-program itself is finite. It's just a finite file of bits. The configuration of the body is the quadruple (body-program *p*, body-input *i*, body-state *s*, body-output *o*). Since all the parts of every body-configuration are finite, every body-configuration is finite. A *body-file* is a digital record of some body-configuration. A body-file is a *perfect physiological ghost*. Of course, a body-file is just the ghost of a 3D body. It is an exact digital picture or snapshot of the body-configuration. The body-file of any earthly human body can be stored on some finite medium like a hard disk. And just as a life is a biologically continuous series of bodies, so a biography is a biologically continuous series of body-files.

Just as every cell has a genotype, so also the body has a genotype. The *genotype of the body* is the genotype of its zygote. After all, for any human body, the zygote is the ancestor of almost every human cell in that body. And the genetic instructions in the zygote direct the construction of the body. The body-program is just the result of multiplying and interconnecting a large plurality of copies of the cell-program encoded by the genotype of the zygote. So the body-program is encoded by the genotype of the zygote. Two bodies run the same body-program if and only if they share the same zygotic genotype. Thus monozygotic twins run the same body-program. Of course, every human body is a visceral ecosystem which includes many cells which are not descendents of its zygote. The body-program is the core of an ecological program, which includes all the programs of all the cells in the visceral ecosystem. But this ecological program is also finite.

29. The life of the body

Every body is a 3D stage of at least one 4D life. A life is a biologically continuous series of bodies. Each next body in any life is the successor of its previous body. Here the logic of the body parallels the logic of the cell. Hence the successor relation among bodies is defined in terms of the successor relation among their configurations. The configuration of any body is the quadruple (body-program *p*, body-input *i*, body-state *s*, body-output *o*). A later body-configuration is a successor of an earlier body-configuration if and only if (1) the later body-program is identical with the earlier body-program; (2) the input to the later body-input is any possible input from the environment; (3) the later body-state and body-output are derived from the earlier body-input and body-state according to some rule in the common body-program.

Figure III.2 illustrates a life with four bodies, each represented by its configuration. The first body is (*p*1, *i*1, *s*1, *o*1) while the last is (*p*4, *i*4, *s*4, *o*4). The vertical lines within each body indicate that its mechanical features occupy the same time. The arrow that runs from each previous body-program to the next indicates that the program remains the same – it just copies itself from body to body. The convergent-divergent arrows show the application of an if-then rule in the body-program. They depict physiological self-updating. For example, from Body-1 to Body-2, the rule specifies that if Body-1 gets input *i*1 while in state *s*1, then Body-2 has state *s*2 and produces output *o*2. For the sake of visual clarity, the positions of the inputs and outputs are swapped from body to body.

Figure III.2 is a connect-the-dots structure in which the dots are mechanical features of bodies and the connections are pipes. The arrows running from each previous body to its next compose an updating pipe. The life in Figure III.2 *spans* many times like a highway spans many places. The connect-the-dots network in Figure III.2 is a 4D, eternal structure. Since there is no identity through time, the bodies in Figure III.2 are all distinct. They are temporal counterparts: every later body in Figure III.2 *was* identical with every earlier body, and every earlier body *will be* identical with every later body.

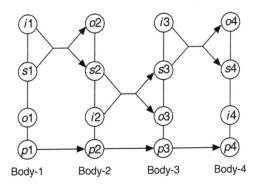

Figure III.2 Four bodies in one life.

30. Digital bodies

A bion is an artificial computer designed for cellular simulation Just as a body is a network of interacting cells, so an *animat* is a network of interacting bions. It is a computer designed for cell-network simulation. Although advanced animats are likely to use specialized hardware, we can treat animats as networks of von Neumann machines.[16] Such networks include the Connection Machine (Hillis & Tucker, 1993) and IBM's Blue Gene (Gara et al., 2005).[17] Of course, the Internet is a network of von Neumann machines. The Internet could be programmed to simulate a network of cells.

Bodies are networks of cells; cells are substrate independent; networks of substrate independent things are also substrate independent; hence bodies are substrate independent. The finite complexity of any body ensures that it is technically possible to make an animat that exactly artificially replicates it. Digitalists therefore endorse a thesis of *functional equivalence* for bodies: for every configuration of any body, it is technically possible to set up an animat with that very same configuration. Thus set up, the configuration of the animat *is identical with* that of the body (of course, the animat is not identical with the body; the animat is merely functionally equivalent to the body). Researchers are building animats that more and more closely approximate entire human bodies (see Section 36). Any animat that replicates some body is a network of bions; but those bions are in turn networks of nanites; hence any animat is ultimately a network of nanites.

Just as any human life is a 4D series of 3D human bodies, so any artificial life is a 4D series of 3D animats. The first body in any life contains exactly one cell (the zygote). As cells divide, earlier bodies change into later bodies. As earlier bodies change into later bodies, so earlier animats change into later animats. As cells are born and die from body to body, bions are activated and deactivated from animat to animat. And as cells make and break connections from body to body, bions make and break connections from animat to animat. The series of animats thus exactly replicates the series of bodies. Since any life is biologically continuous, any artificial life is also biologically continuous.

31. Persons are bodies

A *person* is any thing that can reason and that can behave in a morally responsible way.[18] A person is a rational moral agent. Apart from that, the definition of a person is wide open. Although persons must have certain psychological and moral capacities, the definition does *not* say that persons must be human. There might be nonhuman persons.[19] Other planets in our universe may be inhabited by aliens that can reason and be morally responsible. If such aliens exist, they are nonhuman persons. And even here on earth, engineers might someday create rational moral robots – genuine artificial intelligence. Such robots would be nonhuman persons. But we're not interested in nonhuman persons. We're interested in and only in human persons. A human person is a person with a human body. Unless further qualified, the term *person* means *human person*.

Persons are intimately related to their bodies. *Physicalism* is the thesis that every person is identical with their body.[20] Physicalism is sometimes also called *monism* (to contrast it with Socratic or Cartesian mind–body *dualism*). Physicalism states that if something is a person, then it is a body.[21] Modern science contains a very precise and powerful *Success Argument* for physicalism. It goes as follows: (1) For every function F, if any person can do F, then there is some part of the body of that person whose activity is both necessary and sufficient for the performance of F. All your digestion is done by your guts; all your breathing is done by your lungs; and all your thinking is done by the part of your body that computes. Everything you do is done either by some part of your body or by your whole body (which is an improper part of itself). (2) If everything you do is done by some part of your body, then you *are* your body. Therefore (3) you are your body. This argument is general: every person is identical with their body. Physicalism is successful. Of course, there is a large literature arguing for physicalism (see Churchland, 1985; Dennett, 1993). There is no need to add to it here.

On the basis of our best science, as well as on the basis of much excellent philosophical reasoning, digitalism affirms physicalism. Of course, some people will not accept this – they will want to raise objections to physicalism. But this is the wrong place to deal with them. All the classical digitalists reject Socratic-Cartesian mind–body dualism.[22] For all digitalists, physicalism is axiomatic – and digitalists have every right to start with their own axioms. Digitalism therefore affirms that every person is strictly identical with their body. You *are* your body.[23] Whenever you say "I", you refer to your body. Thus "I am my body" is true every time you say it. And when you say "I am walking" or "I am thinking" or "I am aware" you are referring to your body. Since you are your body, there are many things that you are not. You are not an immaterial thinking substance – you are not a Cartesian mind. There are no such things. You are not any proper part of your body and you are not any property of your body. You are not your brain; you are not your mind; you are not your soul. You are a human person, hence you are a human body. As the existentialists might have put it: *all meaning springs from the body.*

Since digitalists are interested in life after death, they are also interested in experiences which are associated with life after death. Many people try to use *out-of-body experiences* (OBEs) or *near-death experiences* (NDEs) to argue for the survival of an immaterial mind after the death of the body. However, since these experiences have purely physical causes lying within the brain, and since they can be reliably produced by perceptual tricks, or by altering brain functions using drugs or shocks, they provide no evidence for immaterial minds or for the survival of some disembodied mind after death. NDEs are produced by altered brain chemistry or trauma.[24] OBEs are produced by sensory disorientation.[25] They are all just hallucinations. And, like other hallucinations, they are not experiences *of* anything. The extreme vividness of these experiences does not make them true. They do not refute physicalism about persons; on the contrary, since physicalism explains them, they reinforce the thesis that persons are bodies.

Putting physicalism and mechanism together, the result is that *persons are machines.* The argument is easy: (1) every person is identical with their body; (2) every body is a machine; therefore (3) every person is identical with some machine. This is *mechanism about persons.* The idea is old (see Hobbes, 1660; La Mettrie, 1748). And the idea is very specific: *you* are a machine. As a machine, your body is obviously made of matter. But it is matter that is organized according to some form. Your body is highly structured. Every cell has a form; every tissue has a form; every organ has a form. And so your entire body has a form. The form is more than just a shape. It is a dynamic operational pattern. The form of your body is a program for a finitely complex digital computer.

32. Personal impermanence

When you refer to your body, you refer to it *at some time.* When you say "I am my body" at 1pm, you refer to your body at 1pm; when you say "I am my body" at 2pm, you refer to your body at 2pm. But bodies change through time. As bodies change, their parts and properties change. This change refutes identity. Since your body at 1pm and your body at 2pm have different parts and properties, they are distinct bodies. Your body at 1pm *is not identical with* your body at 2pm.[26] There is no identity through time for bodies. As time goes by, one body changes into or turns into a distinct body.

A human life is a biologically continuous process in which all the stages are human bodies (see Section 14). Digitalism says that every person (self) is identical with some body. This means that your self at 1pm is identical with your body at 1pm and your self at 2pm is identical with your body at 2pm. Since those bodies are not identical, those selves are not identical. Your self at 1pm is not identical with your self at 2pm. By affirming that bodies do not remain identical through time, and that persons are bodies, digitalism affirms that persons do not remain identical through time. Persons do not endure; persons do not remain the same through time. Since persons are bodies, and bodies are stages of lives, persons are stages of lives. Persons exist only momentarily – we are not 4D, temporally extended things; we are merely ephemeral and impermanent 3D things. All the persons in any human life are temporal counterparts of one another. Your self at 1pm *will be* your self at 2pm; and your self at 2pm *was* your self at 1pm. Of course, although there is no personal identity through time, there is *personal persistence* through time. Persons persist through time by having temporal counterparts at different times.

Some historical Western philosophers, such as Hume and Nietzsche, denied the identity of the self through time. The denial of personal identity is also found in Buddhism – the Buddhist doctrine of *no-self* (*anatta*) says that there is no permanent person that remains the same through change (Rahula, 1974: ch. 6). All persons (like all things in time) are impermanent. Parfit (1985) is known for affirming something much like this Buddhist doctrine. Although digitalists do not agree with every Buddhist doctrine, they agree with the doctrine of no-self. Rahula very nicely expresses it like this:

> One thing disappears, conditioning the appearance of the next in a series of cause and effect. There is no unchanging substance in them. There is nothing behind them that can be called a permanent Self (*Atman*), individuality, or anything that can in reality be called "I". Every one will agree that neither matter, nor sensation, nor perception, nor any one of those mental activities, nor consciousness can really be called "I". But when these five physical and mental aggregates which are interdependent are working together in combination as a physio-psychological machine, we get the idea of "I". But this is only a false idea.
>
> (1974: 26)

33. The soul is the form of the body

Aristotle said the soul is to the body as form is to matter (*De Anima*, 412a5-414a33). On the basis of this analogy, he declared that *the soul is the form of the body*. This theory was further developed by Aquinas (*Summa Theologica*, Part 1, Q. 75–102). However, Aristotle and Aquinas are very far from modern science. It is often very difficult to make any scientific sense out of their specific soul-theories. Fortunately, there are more scientific ways of thinking about the soul as the form of the body.[27]

Many writers say the soul is to the body as software is to hardware.[28] For these writers, the soul is the form of the body. The digitalist now makes the following argument: (1) your soul is the form of your body; (2) the form of your body is your body-program; therefore (3) *your soul is your body-program*. This definition entails that your soul is the structure (I_B, S_B, O_B, F_B, G_B), and that it is encoded by your genotype. Since your body-configuration contains your body-program, and since your body-program is your soul, your body-configuration contains your soul. Your body-configuration thus has the form (*soul p*, body-input *i*, body-state *s*, body-output *o*). And since your body-file describes your body-configuration, your body-file contains a description of your soul (it also contains a description of your body-input, body-state, and body-output).

For digitalists, the soul is the form of the body. This is an entirely scientific definition of the soul. Here it is important to point out that *your soul is not your mind*. Your mind is an intelligent physical network of nanites; your soul is the form of your body. Your mind thinks. However, since your soul is the form of your body, and since forms don't do anything, your soul does not think. Your soul merely exists as an abstract pattern. Of course, since your soul is the form of your body, it includes the forms of your organs; and your mind is one of your organs; therefore, your soul includes the form of your mind. But the form of your mind is an abstract program that does not think. Digitalists vehemently reject Cartesianism. Souls are not immaterial thinking substances.

Of course, just as your body has a soul, so every network of cells in your body has a soul: the soul of any network of cells is just its program. The soul of your body therefore includes the souls of your mind, your neuro-immune system, your nervous system, and your CNS. Perhaps the soul of your CNS is your *rational soul*. Ultimately, the soul of your body is made of the souls of your cells. This definition of the soul is radically *biological*. If you're used to *psychological* definitions of the soul, then there are at least three ways that the biological definition will seem alien. The first way this definition will seem alien is that (to say it again) *your soul is not your mind*. Your mind is the part of your body that computes. Of course, the form of your body includes the form of your mind (it includes your rational soul). But your soul does not compute. The second way it will seem alien is that it allows distinct people to share the same soul. Monozygotic twins share the same soul. The third way is that your soul may contain errors or mistakes – these correspond to genetic defects. Your soul may be faulty and your soul can be repaired – it can be repaired by genetic therapy.

When your soul is repaired, the result is a new soul. Your new soul is not identical with your old faulty soul. Nevertheless, your new soul still belongs to you – it is one of your body-programs; it is one of the forms of your body. Repairing a defect in your soul is analogous to repairing a spelling or grammatical error in a sentence. Repairing the error reveals the correct meaning of the sentence. Digitalists argue that the correct meaning is *implicit* in the incorrect sentence. Every incorrect sentence implies at least one correct version of itself. Analogously, repairing any genetic defects in your soul reveals its correct meaning. This meaning is part of your *essence*.

Your essence is a system of interrelated souls. Your essence is a set of body-programs which includes your original body-program and is closed under all natural changes. More precisely, a program is in the essence of a body if and only if either it is the program encoded in the ancestral zygote of the body or else it is derived by some natural change from some other program in the essence. Since every body-program contains some mechanisms for self-correction, your body-program naturally entails the correction of every genetic defect.[29] Thus your essence includes all its implied corrections. There are many other types of natural changes of souls – they will be discussed as needed.

Since the soul is the form of the body, the existence of the soul is closely related to the existence of the body. But how closely? Aristotelianism says that forms depend on the things that instantiate them. If there are no things that instantiate some form, then the form does not exist. So if there is no computer that runs your body-program, then your body-program does not exist. Your soul does not exist apart from its physical realizations. If your body is the only computer running your soul, then, when your body dies, your soul ceases to exist. Of course, Aristotelianism is not the only theory of forms.

An alternative to Aristotelianism is *Platonism*.[30] Platonism says that your soul is your body-program; your body-program is the quintuple $(I_B, S_B, O_B, F_B, G_B)$; but each set in that quintuple is equivalent to a set of numbers; thus your body-program is a quintuple of sets of numbers; a quintuple of sets of numbers is a mathematical form; therefore, your soul is a mathematical form. Contrary to Aristotelianism, Platonism says that mathematical forms exist independently of all physical things. Mathematical forms are not parts of any possible physical universes. For digitalists, Platonism is axiomatic.

Platonists are *mathematical realists*. For mathematical realists, mathematical objects (like sets and numbers and programs) are abstract entities that exist independently of all physical things. Mathematical objects are not mental entities. Just as your concept of New York City is not New York City, so your concept of the number two is not the number two. Mathematical objects are not linguistic entities. Just as your name is not you, so the name of the number two is not the number two. Mathematical objects are as real as protons, pulsars, planets, puppies, and people. Since your soul is a mathematical object, this means that your soul exists independently of all its physical realizations.

For digitalists, Platonism is not based on faith. To justify their Platonism, digitalists embrace the *Indispensability Argument* for mathematical objects (Colyvan, 2001). One version of the Indispensability Argument goes like this: (1) Science makes indispensable use of many mathematical theories. (2) If science makes indispensable use of some theory, then that theory is scientifically justified. Hence (3) all the mathematical theories used by science are scientifically justified. (4) But these smaller indispensable mathematical theories are parts of one indivisible all-inclusive mathematical theory. (5) If the smaller theories are scientifically justified, then the all-inclusive mathematical theory is scientifically justified. (6) Therefore, the all-inclusive mathematical theory is scientifically justified. It describes an absolutely rich world of purely mathematical objects. The objects in that world are as real as any other scientifically justified objects (as real as, say, atoms).

34. Naturalism

Over the last few decades, philosophers have developed many definitions of nature which are consistent with our best science. Among these many scientific definitions, digitalists are entirely free to adopt the one that best suits their needs. And so they adopt the one developed by thinkers like Quine (1990) and Post (1999). Their definition of nature is based on the concept of empirical justification.

Following Salmon (1966), digitalists say that a statement is *empirically justified* if and only if it is either a basic statement or the conclusion of some correct inductive or deductive argument from empirically justified statements.[31] Since the outputs of old arguments can be used as the inputs to new arguments, arguments can be stacked to form a rich hierarchy of empirically justified statements. Every statement in this hierarchy is a naturalistic statement, and the hierarchy is the naturalistic hierarchy. A naturalistic theory is any consistent theory composed of statements from the naturalistic hierarchy.

Digitalists are open to any naturalistic theory. And, obviously enough, digitalism itself aims to be naturalistic. Consequently, digitalists need to be more precise about the naturalistic hierarchy. The naturalistic hierarchy has two parameters. The first parameter is the definition of the basic statements. Basic statements include (1) all our most accurate current observation statements and (2) logical statements. Logical statements include the axioms of our best logics (like the predicate calculus). And they include all those *principles of reason* which are used in logic, mathematics, and the sciences.[32]

The second parameter in the definition of the naturalistic hierarchy is the concept of correct argumentation. Naturalism includes all the principles of deductive reasoning. They are well established and can be found in any logic book. Naturalism includes all the principles of inductive reasoning that are indispensable for scientific progress. These principles include *inference to the best explanation,* principles of explanatory and logical closure, and principles of theory *unification* and *simplification.*

More radically, the scientifically indispensable principles of inductive reasoning include some extremely powerful types of *pattern abstraction* and *idealization.* Given these powerful principles of inductive logic, digitalists affirm that the ultimate existence axioms of our best mathematics are inductively derived from observation. Many different systems of ultimate mathematical existence axioms have been defended. Among those systems, digitalists are free to choose the one that best suits their needs. For the digitalism developed here, this system includes the axioms of von Neumann–Gödel–Bernays class theory plus axioms for all consistently definable large cardinals.

Digitalists welcome many arguments that start with observed premises but go far beyond observable conclusions. Such arguments include those for mathematical and merely possible objects. Since the Indispensability Argument is naturalistic, mathematical objects are natural things; yet they are not observable. And since souls are mathematical objects, they are also natural things. Since there are naturalistic arguments for other possible universes, those universes and their contents are natural things. But those merely possible objects are not observable. The old Cosmological and Design Arguments are also naturalistic. If the old Ontological Arguments involve only purely logical statements, then they too are naturalistic. Of course, there is no need to think of these old arguments as arguments for God. Perhaps they are arguments for the existence of other things. There are plenty of new ways to interpret those old arguments.

A *materialist* says that all objects are material; however, since empirically justified theories refer to immaterial objects (such as space-time points), naturalism is bigger than materialism. An *actualist* says that all objects are things in our universe; however, since empirically justified theories refer to non-actual objects (like possible universes and things in them), naturalism is bigger than actualism. Naturalism includes modal realism, which affirms the existence of other possible universes. A *nominalist* says that all objects are concrete; however, since empirically justified theories refer to abstract objects (such as properties, relations, laws, structures, and mathematical objects like numbers and sets), naturalism is bigger than nominalism. The naturalism used by digitalists is a type of Platonism, which includes both concrete objects and abstract objects.

For digitalists, *nature* contains all and only the objects that appear in naturalistic theories.[33] It includes all the objects required by our best sciences. Nature thus includes all the physical things in our universe (such as space-time points, material particles, and wholes composed thereof). Our bodies and minds are among those things. But nature also includes other possible universes and purely mathematical objects. Of course, nature excludes many objects found in obsolete religious, philosophical, and scientific theories.[34] And it excludes the objects found in pseudoscientific theories. Since digitalists are naturalists, the objects excluded by nature are *fictions* – they do not exist.

IV
Uploading

35. Body scanners

A *scanner* takes some thing as its input and produces a description of that thing as its output. As technology has advanced, it has produced a series of better scanners. They can scan ever bigger things ever more accurately. A *body scanner* analyzes an entire body to produce a descriptive *body-file* as its output. Since persons are bodies (Section 31), and since digitalists are interested in persons, digitalists are interested in body scanners.[1]

Familiar techniques (like magnetic resonance imaging, computerized tomography, and positron emission tomography scanners) can be used to imprecisely scan entire bodies. But digitalists are interested in scanners that accurately describe bodies at the level of biomolecular detail. They are interested in scanners that can produce accurate maps of the body at the level of its network of molecular nanites. They are interested in *biomolecular scanners*. Biomolecular scanning is often portrayed as destructive – it *kills* any body that it scans. Perhaps the body is flash frozen and then disassembled molecule by molecule. Since biomolecular scanners have traditionally been parts of hypothetical teleportation systems, they can be referred to as telescanners.[2]

The body-file produced by any biomolecular scanner is a perfect digital image of the body. It is a perfect physiological ghost.[3] Of course, it is the ghost of a 3D body and not of any 4D life. This ghost describes the biomolecular structure of the body. It describes the connect-the-dots network in which the dots are the biologically active nano-machines and the connections are their relations. But this structure is just the configuration of the body. All the biomolecular detail of the body is compressed into a quad of the form (program p, input i, state s,

55

output *o*). Since each item in this quad is just a string of binary digits, the body-file represents four abstract numbers.

36. Digital replication

A bion is an artificial computer designed to simulate a cell (see Section 22). An animat is a network of interacting bions. It is a network of artificial computers designed to simulate a network of cells (see Section 30). It is designed to simulate an entire organism. An animat takes as input some body-file, which it uses to simulate that body. As technology has made progress, it has made a series of ever better animats.[4] An *avatar* is an animat that can simulate all the biologically relevant interactions among all the biologically relevant nano-machines in any organism. It can digitally replicate any organism.

When some body-file is abstracted from some organic body by some telescanner and installed into an avatar, then that avatar digitally replicates that organic body. When an avatar replicates an organic body, it has the full part-whole structure of that body down to the level of biologically active molecules. It contains virtual counterparts of the organs of that organic body. Hence it contains a virtual brain, virtual sense organs, virtual viscera, and virtual musculature. It contains virtual counterparts of tissues and cells. Within the cells, it contains counterparts of organelles. It contains virtual ribosomes and mitochondria. The walls of its virtual cells are studded with virtual receptors. Each virtual cell contains a virtual nucleus with virtual chromosomes and genes. A virtual body has a colony of virtual gut bacteria – it is a virtual ecosystem just like an organic body. Humans are ecosystems. You would not be who you are if you did not have your gut bacteria.

The virtual counterparts in any avatar function in exactly the same biological ways as their organic originals. Thus any avatar is *biologically isomorphic* to its original body – it duplicates its biological activity. When an avatar replicates an organic body, it replicates the network of nanites that performs intelligent information-processing in that body. But that network of nanites is the *mind* of the body (Section 26). Consequently, when an avatar replicates some organic body, then the mind of the avatar replicates the mind of that body. Digital replication is psychologically exact. For example, if some organic body suffers from bipolar disorder, then its avatar will suffer in the same way.

Just as an organic cell has a genotype, so a virtual cell has a genotype; and, just as an organic body has a genotype, so also a virtual body has a genotype. If some organic body has a genetic disease, then its avatar has a virtual version of that disease. For example, if an organic body has cystic fibrosis because of defects in its CFTR genes, and if some avatar digitally replicates that body, then that avatar has analogous defects in its virtual CFTR genes. Avatars can suffer from genetic diseases too. On the bright side, if an organic body has advantageous genes, then its avatars also have those advantages.

An avatar responds to molecular inputs exactly like an organic body. Flavors and odors are molecular. And an avatar has fully realized senses of taste and smell: it can enjoy eating; it can appreciate fine wine; it is aroused by the smell of its lover. An avatar responds to drugs just like an organic body. It has exactly analogous responses to ibuprofen, caffeine, and Prozac. If given virtual LSD, it will begin to hallucinate. It responds to nutrients or their lack just like an organic body. No vitamin C? It gets scurvy. An avatar responds to bacterial or viral inputs just like an organic body. It can be infected with virtual hepatitis or staphylococcus. It can die from the virtual flu. Of course, after any such death, it can simply be restarted in an uninfected state.

37. Terrestrial simulators

A *virtual universe* is a software environment for human habitation (Castronova, 2005). The low degrees of habitability involve little more than the control of low-resolution animats in physically impoverished environments (like Pong, Space Invaders, and Pac-Man). The middle degrees of habitability involve the control of higher-resolution animats in richer environments. The animats in these middle systems more closely resemble human bodies and the environments more closely resemble the earth (consider Second Life, World of Warcraft, and SmallWorlds).[5] As technology progresses, these software environments become ever more humanly habitable. They turn into *terraria*.

A terrarium is a computer system able to simulate at least the terrestrial biosphere at the appropriate levels of detail.[6] The terrestrial biosphere obviously includes the earth. But it includes far more. The terrestrial biosphere is a model of the solar system. It can be pictured as a system of concentric shells like an onion. The innermost shell contains the sun; then there is a shell for every planet (including the earth); there are shells for the outer reaches of the solar system (the Kuiper belt, the Oort Cloud); and, finally, the outermost shell consists of the stars. Most of these shells do not need to be simulated with much detail (Bostrom, 2003: 246–7).[7] Here we focus on the earth.

Most of the resources of any terrarium are focused on simulating the earth. For the purposes of simulation, the earth can also be thought of as a system of concentric shells. The innermost shell is its core; it is surrounded by the mantle and the crust. These geological shells probably do not require detailed simulation (but they cannot be ignored, since they support the magnetic field of the earth). The main part of the biosphere consists of the shells just above and below the surface of the earth. These shells contain the oceans, the land, and the atmosphere. They contain most life, and need to be simulated with great precision – they need to be simulated at the level of biomolecular detail. The outer shells include the higher levels of the atmosphere and the moon.

The purpose of a terrarium is to serve as a virtual habitat for earthly life. Earthly life is not restricted to human animals – it includes the entire earthly ecosystem. Any terrarium must therefore simulate all the features of the planet that are relevant to life. It must simulate the diurnal cycle; the heat and light of the sun; the gravitational forces from the center of the earth as well as the moon; the rotation of the earth; the tilt of the earth on its axis; the magnetic field of the earth; and so on. It must simulate the oceans, the atmosphere, and the weather. It must simulate the dynamics of natural resources. It must simulate the carbon cycle, the oxygen cycle, and the water cycle. It must be biologically realistic.

A terrarium contains an enormous network of avatars. Thus any terrarium can house an enormous civilization – including every human who has ever lived on the earth. But, to realize the entire terrestrial ecosystem, any terrarium also includes avatars for nonhuman organisms. Every terrarium includes animals and plants. It includes fish and birds. It may even include different (and perhaps relatively isolated) regions in which different eras of life are simulated. Any terrarium is biologically exhaustively accurate. It includes all earthly life all the way down to bacteria and viruses – even down to the bacteria that inhabit the cracks in the crust of the earth. Any terrarium boils with life. Obviously, all this virtual life is derived from life on earth. It reconstructs the entire earthly ecosystem. You could be satisfied with less – but why? Since there are no constraints on the powers of possible computers to replicate life, digitalists go all the way to ecological exactitude.

Within the constraints imposed by the need to serve as a habitat for earthly life, there is considerable room for variation. A terrarium need not be the same virtual size as the earth – it may be much larger. Greater size need not imply greater gravity – the density of the virtual earth may be much less. The virtual earth may have a much greater surface area for human settlement. It need not have the same distribution of landmasses or oceans. Terraria may have even stranger features. The moon and other planets can be made habitable. Or perhaps the earth could be replaced with a Niven Ring or a Dyson Sphere. Perhaps there could be different spatial zones corresponding to different periods of earthly evolution. There could indeed be a virtual Jurassic Park or Cambrian Park.

When a terrarium is initialized, it is set up to replicate the earthly ecosystem. It is set up with virtual bacteria; its virtual oceans are filled with plankton and fish; its ponds are populated with aquatic life; its forests are filled with flora and fauna. Of course, many species are impersonal. The differences between the members of an impersonal species are not *morally* valuable (and thus do not *ethically* need to be simulated). Once an ecosystem has been set up with impersonal species, it may then be populated with personal species. Any ethically satisfactory terrarium contains digital replicas of all members of all personal species. Since humans are personal, they must be added to the terrarium one by one. Each human is telescanned and their body-file is *uploaded* into their avatar. Perhaps chimps, dolphins, certain birds, and elephants are also personal species. If so, then they ought to be added to the terrarium one by one, by telescanning and uploading.

38. Teleporting into cyberspace

Uploading involves three devices: a telescanner, some avatars, and a terrarium containing those avatars. All these devices are technologically possible. Digitalism affirms that there are some possible future histories of human civilization that contain these devices. And, within those histories, these devices work as reliably as you please. For the sake of argument, suppose you are living in one of the future histories of human civilization that contain these devices. Human beings upload themselves routinely.

You stand outside of the telescanner. You know that your organic body cannot live much longer (perhaps you are old, or you have a terminal illness, or you know that no matter what you do, you will eventually fall apart). You know that telescanning will destroy your flesh – you will be disintegrated. Finally, you know that telescanning works. If you go into the telescanner and press the green button, your organic body will be digitally replicated in an avatar in some terrarium. On the one hand, you have natural disintegration with no continuity at all: you have *death as extinction*. On the other hand, you have artificial disintegration with virtual continuity: you have *death as disruption*. Faced with this choice, it is rational to choose disruption over extinction. It is rational to choose virtual continuity. Since you are rational, you walk into the telescanner and press the green button. Your body-file is instantly and destructively abstracted from your body.

Your body-file is uploaded into an avatar in a terrarium. Since your avatar replicates your entire body, this is *body uploading*.[8] After your body is uploaded, after it is *teleported* into the terrarium, it becomes a software agent in a software world. The software world is virtual. But there is nothing unreal or immaterial about software. Every computer is a physical thing; every software object supervenes on some computer; but any object that supervenes on a physical thing is also a physical thing; hence your software body is a physical thing, and every object in its software world is a physical thing. Current computers use electricity. Although future computers may use other energies, it can be assumed here that your avatar and its terrarium are both realized electronically. When you upload your body, you become an *electronic body* in an *electronic world*.

Body uploading is *not* disincarnation or disembodiment. It does not imply that some immaterial mind is extracted from your body and that it then persists without a body.[9] On the contrary, body uploading transfers *your entire body-structure* from an organic body to an electronic body. Body uploading does not eliminate your flesh – it merely changes the way that your flesh is realized. After you are uploaded, you do not become disembodied. You have a physical body made out of electricity. Your electronic body metabolizes and is alive (see Alliksaar, 2001). Uploading does not aim to leave the flesh behind; on the contrary, it aims at *the intensification of the flesh*. Body uploading entails that you will persist into a living physical body in a physical environment. Part for part, your virtual body electronically replicates your organic body. Each biologically relevant part is exactly simulated in your electronic flesh. So, even though your electronic body does not have organic DNA, it does have genes and chromosomes. They function exactly as they did in your organic flesh. If your organic body suffered from a genetic defect (such as cystic fibrosis), then so does your electronic body. Your electronic body has cells, organs, and tissues. It has nervous, endocrine, and immune systems.

After uploading, your new electronic body replicates your old organic body down to the biomolecular level of detail. Since the mind is a part of the body, an electronic copy has been made of your mind.[10] Since persons are identical with bodies, this means that an electronic copy has been made of you.[11] Since the soul is the form of the body, the soul of your new electronic body is the same as the soul of your old organic body. Of course, your soul did not move from one body to another. Souls are abstract mathematical objects which do not move at all. Your soul is merely re-instantiated.

Your electronic body now lives in some terrarium. The physicality of your new digital home is sufficiently earthlike for you to live a human life. It is like a video game with realistic physics (for instance, it has 3D space and simulated gravity). Your eyes will see digital light; your ears will hear digital sound; your nose will smell digital odors; your tongue will taste digital flavors; your legs will walk on digital grass under the digital sun; your hands will grasp the digital hands of other uploaded bodies, and your digital brain will experience the sweet release of adrenaline, dopamine, and oxytocin.

39. Processes joined by pipes

A *pipe* is any process that preserves the functionality of a program. A pipe takes an old program as input; it transforms the program in some way that preserves its functionality; and it produces a new program as output. Of course, that transformation may be the identity transformation – a transformation which changes nothing, but produces an equivalent output. Since pipes preserve program functionality, any program that runs into the earlier side of any pipe runs out of the later side of that same pipe. But processes are pipelines. Consequently, if some earlier process is piped into some later process, then one program runs through the earlier process, the pipe, and the later process. Pipes join sequences of shorter processes into longer processes.

One common type of piping is *piping by copying*. For example, a process that was started on a desktop computer can be transferred to a laptop. The transference involves several steps: the program running on the desktop is stopped; the desktop program and its final configuration are copied onto a disk; the contents of the disk are copied onto the laptop computer; the program is started up again in the configuration on the disk. Transference is a pipe. Any program that runs into the transference process on the earlier desktop side runs out of it on the later laptop side. Since transference is a pipe, the desktop process, the transference process, and the laptop process are all one process.

Another type of piping is *piping by porting*. Piping by porting involves copying a program into a new medium (such as a new hardware substrate). Thus porting is a process in which a program running on one type of hardware (such as a Windows PC) is rewritten so that it can run on some other type of hardware (such as a Macintosh PC). True porting exactly replicates the functionality of the original program on the new hardware platform. Of course, in practice, it often happens that programs are not truly ported; they are merely only partially or approximately ported. Nevertheless, since true porting preserves the functionality of the original program, porting is a pipe. Programs run right through ports. A machine has mechanical features, features which compose its configuration (program *p*, input *i*, state *s*, output *o*). When some old machine is ported, its configuration is exactly functionally duplicated in the configuration of the new machine. When one process is ported into another process, the result is a bigger process.

40. One life piped into another

Since lives are processes, one life can be piped into another. A *span* is any series of lives connected by pipes. It is a series of lives in which each previous life is piped into the next. Since pipes link smaller processes into bigger processes, every span is a process composed of lives. Since every span is a process composed of lives, and since every life is a process composed of bodies, every span is a process composed of bodies. This means that all the bodies in a span are processmates.[12] Specifically, they are *spanmates*. Two bodies are spanmates if and only if their lives are in the same span.

Teleportation joins lives together to make a span. Teleportation is an example of piping by copying. For example, teleportation from earth to Mars is a long process involving three smaller processes: (1) the scanning of the earthly body; (2) the transmission of the body-file; and (3) the 3D printing of the Martian body. Since teleportation preserves the functionality of the original body-program, any soul that runs into the teleportation process on the earlier earthly side runs out of it on the later Martian side. The earthly life, plus the teleportation process, plus the Martian life, all make one span. This span is a series of spanmates. These spanmates are the bodies in the earthly life followed by the bodies in the Martian life. And since teleportation is piping by copying, the configuration of the first Martian body is equivalent to the configuration of the last earthly body.

Uploading joins lives together to make a span. Uploading is an example of piping by porting – it's an example of piping by copying into a new medium. If you are uploaded, then your old organic life is piped into a new electronic life. Since uploading preserves the functionality of the organic body-program, any soul that runs into the uploading process on the earlier organic side runs out of it on the later electronic side.

To illustrate how uploading pipes one life into another, consider *Organic* and *Uploaded*. These are two of your lives. Each life is composed

Figure IV.1   One span contains two lives with many bodies.

of three bodies. Organic is the life ⟨You-1, You-2, You-3⟩ while Uploaded is the life ⟨You-4, You-5, You-6⟩. Since You-3 is piped by porting into You-4, You-4 is a digital replica of You-3. When You-3 is piped into You-4, the result is a single span. Figure IV.1 illustrates this span, which has three phases: (1) the original Organic life; (2) the uploading phase; and (3) the Uploaded life in the terrarium. The lives in Figure IV.1 make the span ⟨Organic, Uploaded⟩. Since every life is a series of bodies, every span of lives determines a span of bodies. Every span of bodies runs from the first body in its first life to the last body in its last life. For example, the span of bodies in Figure IV.1 runs from You-1 to You-6.

Figure IV.2 illustrates the logic of uploading in more detail. It depicts a span with four bodies, each of which comes from Figure IV.1. The first body is You-2, which is represented by its configuration (program $p2$, input $i2$, state $s2$, output $o2$). The last body in Figure IV.2

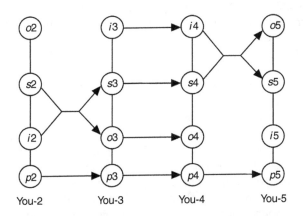

Figure IV.2   Four bodies in one span.

is You-5, which is the configuration (*p*5, *i*5, *s*5, *o*5). For the sake of comparison, the structure of Figure IV.2 deliberately resembles that of Figure III.2 in Chapter III. The arrows that run across the programs in these bodies show that they remain the same: *p*2 is identical with *p*5. The convergent-divergent arrows show the application of an if-then rule in the body-program. They depict physiological updating. But the arrows that run directly from the features of You-3 to those of You-4 show that uploading is porting. Each feature of You-4 is identical with the corresponding feature of You-3 (thus *s*4 is identical with *s*3). Obviously, You-4 is *not* identical with You-3; on the contrary, You-4 is a *copy* of You-3. Copies are distinct things which have identical relevant features.

Figure IV.2 is a connect-the-dots structure in which the connections are pipes and the dots are the mechanical features of bodies. But there are two kinds of pipes in Figure IV.2. These are the updating and porting pipes. Thus You-2 *updates* into You-3; You-3 *ports* into You-4; and You-4 *updates* into You-5. The bodies in Figure IV.2 do not span times. But the *structure* in Figure IV.2 does span times – the arrows span times. The connect-the-dots network in Figure IV.2 is a 4D structure, it is an eternal structure. All the bodies in Figure IV.2 are distinct. Since there is no identity through time, there is neither identity through life nor identity through uploading. So, if uploading differs from life, the difference does not involve identity. Identity is irrelevant.

41. Spans contain many lives

According to Section 14, a process is biologically continuous if and only if every stage in the process is an organism; the process runs some biological program; the stages persist by self-updating; and the process is primarily self-contained. Consider the span composed of Organic and Uploaded. Organic satisfies the requirements for biological continuity. It is a life composed of carbon-based organisms. Uploaded also satisfies the requirements for biological continuity. It is a life composed of electronic organisms.

However, the span composed by Organic and Uploaded is not biologically continuous. Although it satisfies the first and second requirements for biological continuity, it violates the third and fourth. The last stage of Organic does not *update itself* into the first stage of Uploaded. On the contrary, it is *ported* into that first stage – and the porting is done by the *external* agency of the telescanner. Spans are not biologically continuous. Although each life in any span is biologically continuous, the span itself is not. Consequently, while spans contain lives, spans are not lives. No body biologically continues into its digital replica. Both copying and porting *disrupt* biological continuity. But the disruption of biological continuity is death.[13] Within Figure IV.1, the gulf between You-3 and You-4 is death. Uploading kills You-3. You-3 *does not live into* You-4. Since You-3 is not a lifemate of You-4, no Organic bodies are lifemates of any Uploaded bodies.

The positive biological features of spans suggest a type of continuity that is almost but not quite biological. Say a process is *vitally continuous* if and only if every stage in the process is an organism; one biological program runs through the entire process; and every previous organism in the process is piped into the next organism. A vitally continuous process is a series of organisms that run a biological program. Each stage is piped into the next stage; however, the pipe need not be an updating pipe – it might be a copying or porting pipe. For example, the last stage of Organic runs through a porting pipe into the first stage of Uploaded. But porting pipes do not update – they merely duplicate. And when something enters a porting pipe, it does not carry its own form to the other side. On the contrary, its form is carried by some external agency. For example, in uploading, the form of You-3 is carried by the telescanner into You-4. Spans that include copying or porting are vitally but not biologically continuous. The difference between life and uploading does not involve identity – it involves continuity. Ordinary lives are continuous in one way while spans are continuous in another way.

Spans that include porting are vitally continuous. Vital continuity entails persistence: within any span, every previous stage *persists into* its next stage. On the one hand, vital continuity does *not* give you life *through* death. Hence uploading does not give you life through death. Life through death is impossible; if you want life through death, then you want something impossible; if you want something impossible, then your desire is self-defeating. Your desire is based on an attachment to presence that can only breed suffering. On the other hand, vital continuity gives you life *after* death. Hence uploading gives you life after death. If you are uploaded, then you die; but after you die, you will live again. Uploading enables you to *prevail over* your death and to *overcome* it.

42. Stages of spans are counterparts

A process is a series of stages that runs a program. Since the span containing Organic and Uploaded runs a soul, it is a process. It is a span composed of two lives. All the bodies in that span are spanmates. Since they exist at different times, they stand to one another in temporal relations. Within any span, one body may be earlier than another or later than another. And every body in any span is simultaneous with (and only with) itself. Within any span, any earlier body is a *past spanmate* of any later body, any later body is a *future spanmate* of any earlier body, and every body is a *present spanmate* of itself. Since spans are processes, all the bodies in any span are temporal counterparts. They serve in the truth-conditions of tensed statements about the past, present, and future.

The Past. Any statement you make about yourself in the past tense is made true or false by your past spanmates (your past counterparts). For any predicate P, any statement of the form "I was P" is true when said by you now if and only if one of your past spanmates is P. Consider how this works when the spanmate relation crosses two distinct lives. Consider what some stage of Uploaded might say about some stage of Organic. Thus "I used to be an organic body" is true when said by You-4 if and only if it has some past spanmate who is an organic body. And it does have such a past spanmate: You-2. And, since each body in Uploaded *was* identical with each body in Organic, each body in Uploaded can truly say: "I was identical with an organic body".

The Present. Any statement you make about yourself in the present tense is made true or false by your present spanmate. For any predicate P, any statement of the form "I am P" is true when said by you now if and only if your present spanmate is P. For the present tense, temporal counterpart theory is trivial – every body is its own present spanmate. Thus "I am ok" is true when said by You-2 if and only if it is ok, which it is.

The Future. Any statement you make about yourself in the future tense is made true or false by your future spanmates (your future counterparts). For any predicate P, any statement of the form "I will P" is true when said by you now if and only if one of your future spanmates is P. Consider again how this works across lives. Consider what some stage of Organic might say about some stage of Uploaded. Thus "After my organic body dies, I will live again as a new electronic body" is true when said by You-2 if and only if it has some future spanmate who is an electronic body. Fortunately, You-2 has the future spanmate You-4. Hence You-2 will live again as an electronic body. And, since each body in Organic *will be* identical with each body in Uploaded, each body in Organic can truly say: "I will be identical with an electronic body."

43. Uploading is resurrection

Uploading is not the disincarnation and reincarnation of some immaterial mind; on the contrary, it is the destruction of an organic body and the construction of an equivalent electronic body. It is the replication of the old flesh in a new digital medium.

As such, uploading very closely resembles the *replication theory of resurrection* developed by the philosopher John Hick (1976: ch. 15). According to Hick, resurrection goes like this: (1) The last stage of some earthly life is *Fallen*. (2) When Fallen dies, a replica of Fallen is created in some other universe. (3) This replica of Fallen is *Risen*. Many other writers have given similar definitions of resurrection.[14] Consequently, it is plausible to say that the construction of a digital replica of some organic body is the resurrection of that body. This is *resurrection as digital replication*.

The definition of resurrection looks like this: an earlier body *will be resurrected* if and only if it *is* an organic body and it *will be* a resurrection body.[15] But what does it mean to say that an earlier body will be a resurrection body? Within the context of uploading, resurrection bodies are electronic. Thus an earlier body will be resurrected if and only if it is an organic body and it will be an electronic body (produced by uploading). Temporal counterpart theory tells us that an earlier body will be an electronic body if and only if it has some future electronic spanmate. Its future electronic spanmates are its *resurrection counterparts*. Putting this all together, an earlier body will be resurrected if and only if it is an organic body and it has some resurrection counterpart.

Figure IV.3 illustrates the concept of digital resurrection. Arrows with open heads indicate biological continuity. The last living stage of your organic life is *Fallen*. Fallen is You-3 in Figure IV.1. Fallen dies during

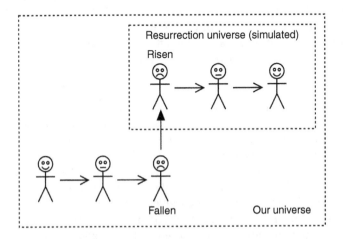

Figure IV.3   From our universe to a simulated resurrection universe.

the body uploading. The first living stage of your new electronic life is *Risen*. Risen is You-4 in Figure IV.1. Risen is brought to life during the body uploading. The arrow with the solid head, which goes from Fallen to Risen, indicates that Risen is a digital replica of Fallen. Every lower organic body in Figure IV.3 *will be* resurrected in every upper electronic body.

On this definition of resurrection, the statement "I will be resurrected" is true when said by you now if and only if you have a resurrection counterpart. Likewise "I will be identical with a resurrection person" is also true when said by you now if and only if you have a resurrection counterpart. And Fallen can say truly that they *will be resurrected* and *will be identical* with a resurrection body. Tense matters. Since Fallen is an organic body now, Fallen cannot truly say that they *are* identical with a resurrection body. For any resurrection body, a statement like "I did the bad (or good) deed on earth" is truly said if and only if it has a past spanmate who did the bad (or good) deed. So Risen can truly say "I did the deed on earth" if and only if some past spanmate did it.

44. Resurrection is self-actualization

You walk into the telescanner and press the green button. When you press it, you are instantly killed. Of course, after you die, your avatar appears in its terrarium. You have been uploaded. Your organic life is followed by your new uploaded life. The span that contains those two lives is not biologically continuous. But so what? The purpose of resurrection is not to ensure the continuity of your old organic life. On the contrary, the purpose of resurrection is to ensure the further realization of the positive potentialities of your soul. Resurrection does not give you life through death; it gives you life after death. Resurrection gives you a *new life* in a *new environment*. The purpose of your resurrection is to *overcome* or to *surpass* the defects of your organic life.

For example, suppose Fallen is a man with a defective CFTR gene. This gene defines part of the logic of his body-program – it defines part of the logic of his soul. From the very start, from his zygotic beginning, that logic is defective. The errors in just a few molecules at his origin determine his entire system of life-potentials. And, from the perspective of biology, those potentials are not very good. Fallen has *cystic fibrosis*. The life of Fallen will be short and filled with suffering. Of course, the potentials of Fallen are not defined *entirely* by his broken CFTR gene. The genome of Fallen encodes an entire system of healthy human potentials (the potential for a long life; the potential to be a great athlete). But the broken CFTR gene overrides those potentials. To actualize them, that gene, and any other damage associated with it, must be repaired.

The purpose of resurrection is not to ensure the continuity of a diseased life. It is not to ensure the continuous operation of a defective body-program. On the contrary, the purpose of resurrection is to *liberate* the healthy potentials of that soul from its diseased potentials. The purpose of resurrection is to glorify the flesh, to spiritualize the flesh, to intensify the visceral excellence of living computation. When Fallen is resurrected into Risen, Risen inherits the soul of Fallen. Risen is not identical with Fallen; but the soul of Risen is identical with the soul of Fallen. Risen does not have the DNA of Fallen; but Risen does have the genes of Fallen. Thus Risen inherits all the potentials of Fallen, including both the healthy and the diseased potentials. Risen also has cystic fibrosis.

At least one of the purposes of uploading is to provide a context in which the defects of body-programs are more easily corrected. Digital genes are easier to edit than carbon-based genes. After Fallen is resurrected into Risen, Risen will be restored to health at the genetic level. Every broken CFTR gene in every cell in Risen will be *corrected* so that its meaning becomes clear. Of course, since the correction of a defective body-program preserves its functionality, correction is a biological pipe. The life of Risen continues through correction. Once the cystic fibrosis is cured, many of the healthy potentials of Risen are liberated from their organic shackles. But Risen inherited those healthy potentials from Fallen. Hence the healthy potentials of *Fallen* will be liberated from their organic shackles. Neither biological nor psychological continuities are required for this liberation. On the contrary, those continuities frustrate that liberation. Rather than continuing the old defective life of Fallen, Risen now goes on to lead a better kind of life in the terrarium. Risen is free to become a great athlete and to lead a long and happy electronic life.

As Risen goes on to lead that better life, Risen is actualizing the potentials that lay hidden in Fallen. Risen is doing something positive for Fallen, namely, actualizing his implicit positive potentials. And Fallen *ought* to appreciate this. As he presses the green button, Fallen ought to be thankful for the opportunity to be uploaded; Fallen ought to be thankful for the future existence of Risen. And Fallen ought to be thankful that Risen will *not continue* his life. Fallen ought to be thankful that Risen will *improve* his life. Thanks to Risen, Fallen has properties that he otherwise would not have had.[16] These are future-directed properties. Thanks to Risen, Fallen can truly say: "I will be cured of my disease;

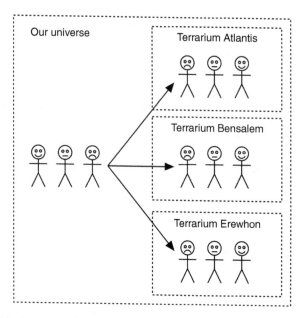

Figure IV.4   An organic life is multiply uploaded.

I will live a better life." Of course, even if you don't have cystic fibrosis, there are surely many ways that uploading can improve your life.

There are many ways to improve (the rest of) any life. If you are uploaded, then one of those ways will be actualized. Of course, if it is good to actualize one of those ways, then it is better to actualize many of those ways, and it is best of all to actualize every way to improve your life. Uploading provides you with the opportunity to actualize many ways to improve your life. You can be uploaded many times – or, more precisely, many copies of your last organic body can be distributed into many terraria. Each terrarium will provide you with some different circumstances in which your different positive potentials can flourish. Across these different terraria, your soul will continue to unfold in different ways. Figure IV.4 illustrates multiple uploading. The terraria in that figure are named for various fictional utopias. Multiple uploading is *body-fission* (Section 13). The spans in Figure IV.4 overlap – they all share the same initial organic stages.

45. Uploading into utopia

After you are uploaded, your replica appears in the resurrection universe. This resurrection universe is a terrarium – it is an artificial universe *intelligently designed* by some possible future human engineers to be a habitat for humans. When they devise any terrarium, those engineers are engaged in *universe-design* (also known as *world-design* or *world-building*). Universe-design is a kind of large-scale architecture. It is a well-established discipline in many creative arts. The techniques of universe-design are used by science fiction writers, futurists, and digital game designers.[17]

One of the major premises of uploading is that the engineers design their terraria to be habitats for human flourishing. Terraria are not built for torture or vengeful punishment; they are built to be heavens rather than hells. This premise is justified by the *Argument for Virtuous Engineers:* the engineers have built sophisticated technologies; building such technologies requires long-term stability and rational purposiveness; but those qualities require social harmony; social harmony requires virtuous individuals virtuously organized; hence the engineers are virtuous persons in a virtuous society; but virtuous persons will design habitats for the flourishing of other persons. Much of what ought to be true on earth is not; but more of what ought to be true on earth will be true in terraria.

Since *utopias* are traditionally said to be places that are located within our universe and that are intelligently designed to be more congenial to all that is best in humanity, the terrarium is a *digital utopia* (Coenen, 2007). Many writers have defined utopian places. Since terraria are self-contained, they are like *island utopia* (Clay & Purvis, 1999). Perhaps some would be like Plato's Atlantis or Bacon's Bensalem.

46. Designing utopian physics

Since any utopian terrarium (any resurrection universe) is designed to facilitate human flourishing, it must be very similar to its surrounding organic universe. Its physics and chemistry must be very similar to their organic versions. It must have three spatial and one temporal dimension; it must have natural laws; it must have relatively stable material things. Your new uploaded body will live on the surface of an earthlike planet with an earthlike atmosphere and earthlike gravity. Your new habitat will be illuminated by a sunlike star and moonlike moon. Since you have a body, you will interact bodily with your environment. You will breathe digital air, drink digital water, and eat digital food.

Any habitat for humans must be governed by laws. As rational animals, our lives require a rationally comprehensible environment. Even video games have rules. If an environment does not have rules, then we cannot learn how to live in it – we can't learn to operate it. An environment without rules would be unpredictable; its unpredictability

would lead to anxiety and depression. Nobody would be able to rationally assess risk or danger. Nobody would ever undertake any project.
An environment without rules would be unintelligible; its unintelligi-
bility would lead to psychosis. It is easy to imagine having the powers
of cartoon or comic book superheroes. But it is difficult to design a rule-
governed universe in which those powers are coherently possessed by
human bodies.

For example, consider the power of flight: it is easy to *imagine* that
you can fly anywhere in the afterlife. But how would flying be *realized*
in a universe that serves as a habitat for replicas of organic bodies? Many
laws of nature would have to be *broken* for a human body to fly. How
would we steer or speed up or slow down? Our bodies are not like avian
bodies – we don't have wings or tails. Would it require physiological
effort to fly? Would you feel tired after flying? How many calories would
be burned by an hour of flight? Would you have to get into shape for
flying as you do for running? What about the dangers of falling or col-
liding? The ability to fly must make physical and physiological sense.
If it does not, then it is little more than a meaningless hallucination.

Since utopia is governed by physical laws, success in the afterlife
requires physical effort and work; you do not accomplish anything in
the afterlife merely by wishing it or willing it to be so. Utopia is not a
fantasy land.[18] It is not a *magical* place. It is not a world in which your
will can work *miracles.* The fact that the will is not magically effective
is required for any meaningful human life. For if the will were magically effective, our uncensored thoughts would cause effects, and the
result would be chaos. Human impulses to sex and violence would produce nightmares of hellish conflict. The fact that it takes effort to do
something is what filters out worthless spontaneity. And the worthiness
of any goal is directly proportional to the effort needed to realize it.
Our projects are meaningful and valuable in direct proportion to their
difficulty. And so it must be in the afterlife: meaningful and valuable
accomplishments will be the result of hard work.

47. Designing utopian biology

Since life in utopia is life in the body, the foundation of utopian life
is a healthy body. It's almost impossible to find a human in perfect
physiological condition. Almost nobody is in perfect health. And it's
reasonable to think that, if you're being uploaded, then you're near the
end of your earthly life. You're probably old and sick. Perhaps you've
got cancer or Parkinson's disease, or some other disease of age. If you've

been diagnosed with the early stages of Alzheimer's, then you'll want to get uploaded before it really begins to damage your brain. But you might get uploaded when you're young. You might suffer from a horribly painful condition. Perhaps you have multiple sclerosis. Or you might suffer from a genetic disease like cystic fibrosis. Or maybe you were injured in an accident or in war. Perhaps you have lost a limb, or you are blind or deaf.

If you aren't perfectly healthy when you get uploaded, then your electronic body will be *healed* in utopia. And if you aren't in the prime of life when you get uploaded, then your electronic body will be *rejuvenated* to bring it into a state of optimal youthful functionality. On these points, uploading parallels traditional resurrection theories, which state that your resurrection body will be healed and rejuvenated (Hick, 1976: 294; Tipler, 1995: 242–4). Fortunately, it is relatively easy to modify digital flesh (indeed, since digital flesh is designed for human flourishing, it is much easier to repair and enhance than organic flesh). It's a matter of debugging your body-program and fixing errors in your body-state. For example, digital medical procedures scan your electronic body for cancer. Every cancer cell is corrected at the molecular level. If you are missing limbs, they are regrown from your genotype. If you are blind, your retinas or optic nerves are repaired so that you can see. And if you have cataracts, your lenses can be repaired. Generally speaking, the intelligibility and editability of digital flesh allows all bodily defects to be corrected.

The healing and rejuvenation processes are continuous rather than disruptive. They gradually construct your new digital flesh. For instance, if you are obese, you gradually lose weight and gain muscular strength; if you are bald, your hair gradually regrows. These changes occur at a natural rate that preserves biological continuity. Any diseases contracted in utopia will always cure; any injuries suffered in utopia will always heal. And there will be no permanent death or extinction in utopia. Of course, this does not mean that you are invulnerable. On the contrary, there is real risk in utopia – you may die. But you always have a backup version of yourself (Moravec, 1988: 112).[19]

Life in utopia is life in the body. Since your new electronic body remains human, it replicates all the *essential* features of your old human physiology. It has your human biological nature. Your electronic body breathes, drinks, and eats. It has all the normal human organs, which perform their natural functions. Your electronic body will have a diurnal cycle: it will go to sleep and wake up. You may sleep less and be more energetic. But humans are not naturally awake all the time. Since sex is a natural part of human life, you will be able to have sex in utopia (Tipler, 1995: 256–7). You may even produce children and raise a family. Unrestricted by scarce resources, and unconstrained by disease, sexual relations in utopia may be quite intricate. Marriage may be complex.

48. Designing utopian society

Since humans are social animals, and since utopia is designed for human flourishing, you will not be in utopia alone. You will live in a utopian society. And, just as the terrarium is built within the earthly universe, so the utopian society is built within some earthly society and therefore within some earthly legal system. The earthly legal system will define the legal relations between earthly and uploaded people. It will decide whether uploaded people can own earthly property; can be married to earthly people; can vote in earthly elections; can sue in earthly courts; and can exercise political powers over earthly people.

It is plausible to say that a society is *optimal* if and only if it is managed in a way that aims to maximize both the happiness of every individual and the justice within every collective. And a society is *self-optimizing* if and only if it naturally tends towards optimality. It is arguable that human societies are not naturally self-optimizing. We are not able to direct ourselves towards what is best for ourselves. If that is right, then any optimal society requires some nonhuman manager. Since utopia is intended to be an optimal society, this means that it requires some nonhuman manager. Some *superhuman* and *superethical* managerial intelligence is needed to optimize the relations among humans in any terrarium. It is plausible that any civilization sufficiently advanced to carry out large-scale uploading will also be sufficiently advanced to design artificial intellects that are both cognitively and ethically superior to humans. Call them *angelic artillects*. This line of reasoning says that utopia will be managed by angelic artillects.[20] These angelic artillects are software agents running at the subphysical levels of their terraria.

Since life in utopia will be meaningful, it will also be challenging. Perhaps it will be like a game in which you work through levels of progressively more difficult adventures. Or perhaps you will have the opportunity to realize more and more of your talents through training and practice. It will be just as hard to achieve greatness in utopia as it is to achieve greatness on earth. Attempting to achieve greatness, you will fail many times. But if you work at it long enough and hard enough, you will achieve it. You will write a great novel; be an amazing lover; win an international athletic contest. Both the failures and successes in utopia are real. Failure will really hurt and success will really feel good. And, like on earth, your choices are embedded in networks of antecedents and consequents. If you hike into a jungle, and you decide you want to go home, you will have to hike back out. The risks are real: if you climb a mountain, you may become injured or fall to your death. You can suffer and die. But failure and death in utopia are never permanent. If you get injured, you will always heal. If you die, your last archival self is brought to life again.

49. Beyond uploading

After you are uploaded, your replica appears in the resurrection universe. This resurrection universe is a terrarium – it is an enormous computer programmed to simulate the appropriate parts of our universe in the appropriate ways. As a hardware platform on which something like our universe runs, the terrarium resembles the deity defined by Spinoza. The terrarium is something like a divine computation in which the uploaded people *live, move, and have their being* (Acts 17: 28). It continuously creates these people, sustaining them in their being by its constant self-updating. Would it be appropriate to refer to the terrarium as a *god*? Why not? The terrarium is already far more glorious than many old-fashioned gods. The terrarium is extremely powerful. And, through its angelic artilects, it is extremely intelligent and benevolent.

You might object that the terrarium can't be a god because it was designed and created by some possible future humans. On this view, if there are any gods associated with uploading, they are the human engineers who designed and created the terrarium. After all, since the terrarium is a kind of universe, its designer-creators are like the designer-creators of universes, and it is plausible to say that such designer-creators are gods. Moreover, the engineers who made the terrarium have power over it – they can turn it off. Presumably, if you are living in a terrarium, then you remember your past life, and you probably don't think of the engineers as gods. After all, they are human-all-too-human, just like you. But after a long time in the terrarium, you might forget. And if people have children in the terrarium, then somewhere along the line, your descendents might forget. The truths about the gods might be preserved only in very inaccurate ancient stories.

All this suggests that we may already be living in a terrarium. If it is possible for future humans to build a terrarium, then it is also possible that we are already living in that future. Our ancestors built this terrarium long ago; some of them uploaded themselves into it, and we are their descendents (or perhaps we are repeated versions of humans who lived long ago). Alternatively, it may be that our universe is a terrarium built by some superhuman engineers. We might already be living in a software universe. Its engineers would be like gods to us. Uploading suggests the deeper possibility that the entire *physical* structure of our universe is embedded in a much larger *natural* structure. It suggests that our universe is one small part of nature. But this possibility is the topic of *promotion*.

V
Promotion

50. Digital physics

Many arguments reason from natural evidence to the existence of some *necessary ground* of our universe. This ground is the creator and sustainer of all the physical things in our universe. Arguments for this necessary ground of our physicality include Aquinas's Second and Third Ways (*Summa Theologica*, Part 1, Q. 2, Art. 3).[1] These arguments are usually referred to as *cosmological arguments*. Since they are based on natural evidence, cosmological arguments are naturalistic arguments. If they are sound, then they produce natural outputs – they cannot lead to supernatural objects.

The *Argument for our Ground* is a kind of cosmological argument. It goes something like this: (1) Our universe contains many higher-level objects supported by societies of lower-level objects. Organisms are supported by cells; cells are supported by molecules; molecules by atoms; and atoms by particles. And, since our best physics says that particles are just ripples in space-time, particles are supported by space-time points. (2) There are many descending support chains (in which x's rest on y's, y's rest on z's, and so on). (3) Support chains have no loops. (4) Support chains cannot be infinitely descending.[2] (5) Thus, after finitely many steps, every support chain bottoms out in some independent things at the lowest level. (6) All dependency chains starting in the same universe bottom out in the same system of independent things. (7) Since any universe is unified, the system of independent things at the bottom is also unified. Those independent things constitute a whole at the lowest level. (8) This whole is the necessarily existing ground of our universe. It is the creator and sustainer of all higher-level things in our universe. This whole exists beneath the level of space-time points; it supports those points. Assuming that points are the deepest *physical* things, the ground of our universe is *sub-physical*.

Although digitalists accept the Argument for our Ground, the nature of our ground remains to be clarified. Here digitalists turn to work on the computational foundations of physics. According to *digital physics*, our universe is a software process running on some computer (Zuse, 1969; Steinhart, 1998). Our universe is virtual. Of course, virtuality does not imply that our universe is not real. It just means that it is not *ultimately* real. Just as a wave supervenes on some water, so all the physical things in our universe supervene on some computer. This computer is the *Engine* (Fredkin, 1992).

Digital physics is an abstract thesis about the complexity of our universe. It entails that our universe is computable. It *might* be computable by a classical Turing machine (Turing, 1936; Hopcroft, 1984). However, digital physics *does not* require that our universe is classically Turing computable. After all, *there are many kinds of computers that surpass classical Turing machines.*[3] Our universe might be running on some infinitely complex super-Turing computer. And digital physics *does not* entail that our universe is a cellular automaton. Digitalists use cellular automata as convenient models because they are easy to understand; but there are other ways to do digital physics.

Four lines of reasoning support digital physics. The first line is theoretical. Many physicists have argued that our universe is ultimately computable (Fredkin, Landauer, & Toffoli, 1982; Deutsch, 1985; Fredkin, 1991; Zeilinger, 1999; Fredkin, 2003). And many computer scientists have made similar arguments (Schmidhuber, 1997; Wolfram, 2002). The second line is based on digital simulations of our universe. Highly accurate simulations of our universe have been made at both the large scales of relativity theory (Springel, 2005) and at the high levels of the resolution of quantum mechanics (Gattringer & Lang, 2009). Future refinements of those simulations may converge, at least in principle, to exactitude. The third line comes from studies of the informational capacity of our universe. Lloyd (2002) calculates that our universe so far has run only 10^{120} operations involving 10^{90} bits. These are finite numbers. If our universe is finitely complex, then it is computable. The fourth line is statistical. Zenil & Delahaye (2010) have developed statistical methods for testing the hypothesis that processes in our universe are algorithmically generated. They discovered correlations that support that hypothesis. If all the processes in our universe are algorithmic, then our universe is computable.

There are some immediate objections to digital physics. The first is that the apparent continuity of current physics conflicts with the apparent discreteness of digital physics. The reply is that it is indeed fair to present this as a conflict between appearances. On the one hand, the continuities of current physics may turn out to be over-idealizations. On the other hand, the discreteness of digital physics is an under-idealization. There are many ways to extend the concept of computation beyond classical discrete computation. The Engine may perform continuous computation on real numbers (Moore, 1996; Blum et al., 1998). Our universe may be the limit of an infinite series of increasingly complex and detailed finite computations. The second objection to digital physics is that there is no available reconstruction of current physics in computational terms. The reply is that many scientists are working on digital reconstructions of current physics.

51. The Engineers

One of the more popular interpretations of digitalism says that the Engine was designed and created by some *Engineers* (Moravec, 1988: 122–4, 152–4, 178; Moravec, 1992; Tipler, 1995; Bostrom, 2003; Chalmers, 2005). This interpretation of digitalism is *simulism*. Two arguments provide simulism with some empirical justification.

Simulism gains empirical support from the *Simulation Argument* (Bostrom, 2003; see also Weatherson, 2003; Bostrom, 2005; Brueckner, 2008). There are many versions of the Simulation Argument. One version goes like this: (1) It is highly likely that every civilization like ours will mature further into one that is able to run vast numbers of universe simulations. (2) It is highly likely that any civilization able to run vast numbers of universe simulations will want to run them. (3) Any civilization that can run simulations and that wants to run them will run them. The engineers in those civilizations will therefore design and create engines that run universes. (4) Minds realized by computers are psychologically indiscernible from minds realized by flesh. Consequently, (5) there are enormously many more simulated minds than nonsimulated minds. (6) Which makes it highly probable that we are simulated minds living in a simulated universe. Our universe is running on some Engine that was designed and created by some Engineers.[4]

Simulism also gains some empirical support from the *Fine Tuning Argument*.[5] It goes like this: (1) Our universe is running on the Engine. (2) Whatever else the Engine may be, it is a computing machine that is running a universe-generating program. Since the nature of our universe is encoded in that program, it can be referred to as our *cosmic script*. (3) Our cosmic script does not have to be the way that it happens to be. On the contrary, there is some vast library of possible universe-generating programs that the Engine can run. (4) Since there are many such programs, there has to be some explanation for why the Engine runs our cosmic script. (5) The Engine is analogous to an earthly computer. (6) If some earthly computer runs some specific program rather than some others, the best explanation for that fact is that the specific program was developed by some software engineers in order to serve some purpose. (7) By analogy, the best explanation for the fact that the Engine runs our cosmic script is that it was developed by some Software Engineers in order to serve some purpose. As it was developed, the features of our cosmic script were *finely tuned* to serve that purpose. Continuing the analogy, the best explanation for the fact that the Engine exists at all is that it was developed by some Hardware Engineers in order to serve the purpose of the Software Engineers. (8) These Software and Hardware Engineers together are the *Engineers*. Consequently, the Engineers exist.

The *Argument from Special Features* now reasons from the finely tuned features of our universe to the purpose of the Engineers: (1) Our universe has certain finely tuned features. These include (a) the fact that our universe is lawfully ordered; (b) the fact that the laws of nature have certain mathematical forms; (c) the fact that certain special constants appear in those laws; and (d) the fact that the universe starts in a low-entropy state. (2) These features are significant because if they were slightly different, then our universe would contain no things that have values. The values include complexity, life, intelligence, rationality, personhood, and so on. (3) All these features are encoded in our cosmic script. (4) The best explanation for the choice of our cosmic script is that the Engineers wanted to make a universe filled with things that have those values. They wanted our universe to be filled with things that are complex, living, intelligent, rational, personal, and so on. (5) And the best explanation for that desire is that the Engineers themselves are benevolent, intelligent, and powerful. To paraphrase Leibniz (*Monadology*, 55), they *want* to make a universe filled with value; they *know how* to make a universe filled with value; and they *can* make a universe filled with value; therefore, they do. They make that universe by building the Engine and then setting it up to run our cosmic script.

The *Proportionality Argument* further clarifies the nature of the Engineers: (1) The Engineers designed and created our universe. (2) As Hume says, the perfection of the designer is proportional to the perfection of the designed (1779: Part 5). (3) Since our universe is only finitely perfect, the Engineers are only finitely perfect. (4) Therefore, the Engineers are only finitely perfect.[6] So, who are the Engineers? An intriguing answer is provided by the New Atheists. Dawkins discusses the simulist hypothesis that our universe is running on a computer in some other universe (2008: 98–9, 184–9). And Harris says: "If intelligently designed, our universe could be running as a simulation on an alien supercomputer" (2008: 73). Digitalism strives for naturalism. So, the Engineers are physical persons in some other universe. To say that they are persons does *not* mean that they are humans – it means only that they are rational moral agents.

Although the Engineers are only finitely perfect, that finiteness does not prevent them from being *divine*. For digitalism, something is divine if and only if it is a cosmic designer-creator. It is divine if and only if it realizes the best universe(s) in some set of available alternative universe-forms. However, the Engineers are not supernatural spirits. On the contrary, they are physical entities operating a physical machine that makes little universes inside their big universe. The digitalist says the Engineers are *naturalistic divinities*. Figure V.1 illustrates the way our universe is nested in its higher universe.

52. The higher universe

The Engineers live in their own higher universe. What is the physical nature of this higher universe? How does it work? There are at least three good reasons to think that the physics of the higher universe is similar to our physics.

The first reason is that our universe is running inside of the physics of the higher universe. It can deviate only in ways that are permitted by the higher physics. Greater deviations are harder to compute; so it

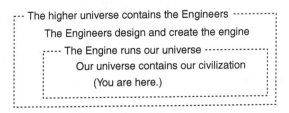

Figure V.1 The Engineers in their higher universe.

is reasonable to think that the physics of our universe is made in the image of the physics of the higher universe.

The second reason is that when we design simulations (such as video games), they have physics similar to ours. If they did not, then we would find it difficult to play those games. The similarity often involves simplifications of our physics (we live in a 3D space, but the video game is a 2D space). The similarity often permits *easily intelligible* variations of our physics (people can fly). And easily intelligible variations are easily computable variations – they are small rule-governed variations. Hence those variations stay close to the physics of our universe. Video games don't add a fourth spatial dimension; nor do they add novel forces or types of matter that are governed by hard-to-compute mathematical laws. Thus physics is made in the image of the higher physics.

The third reason is that we are social. We are interested in humans or in things like humans. So we make video games in which humans or humanoids can operate. It is reasonable to think that advanced intelligence goes with advanced sociality. Hence the Engineers are social; hence they are interested in things like themselves; hence we are made in their images and our physics is made in the image of their physics. Of course, this does not mean that they are humans. We may be highly simplified versions of them. As other apes and monkeys seem to us, so we may seem to them. To the Engineers, our levels of intelligence and morality are likely to seem primitive and crude indeed.

Since the Engineers have designed and built our universe, it is clear that they collectively possess superhuman intelligence and superhuman power. But the Argument for Virtuous Engineers (Section 45) justifies the thesis that they also collectively possess superhuman benevolence. The Engineers are like angels. Their higher civilization, which exists in the higher universe, is better than any merely human civilization.

53. The simulation hypothesis

Many software universes involve cellular automata, like the game of life (see Section 9). It is possible to construct a universal Turing machine in the game of life (Rendell, 2002). And it is also possible to construct self-replicating patterns – simple life forms (Poundstone, 1985: ch. 12). Since mentality is computable, the game of life can support living thinking patterns. These may be as smart as human animals.

One of the classical digitalists, Hans Moravec, uses the game of life to tell the story of *Newway* and the *Cellticks* (1988: 152–3). Newway is an engineer in our universe who runs a game of life universe on some computer in his laboratory. As he experiments with the evolution of ever more complex patterns, some of them become intelligent. These are the Cellticks. The Cellticks learn to communicate with Newway. He tells them about our universe and invites them to join it. Since the Cellticks are intelligent programs, and since intelligent programs can control robotic bodies, Newway builds robotic bodies for the Cellticks. These robotic bodies are things in our universe, outside the computer. Newway then transfers the Cellticks into these robots. The Cellticks evolved inside a software universe and then graduated – they became embodied in the surrounding hardware universe. Each Celltick got *promoted* from its universe into ours. Promotion entails persistence: every Celltick persists into its new robotic body.

Although Moravec's tale of Newway and the Cellticks is fiction, something similar has been done in fact. Lipson and Pollack (2000) designed some simple software robots. Their bots were composed of four types of software building blocks: bars, mechanical actuators, ball-and-socket joints, and neurons. These bots were evolved in a purely electronic world with 2D space (a flat plane). They were evolved to move (the evolutionary fitness measure was mobility). After many generations, bots evolved that moved around on the plane. Some of the fittest bots were then selected for promotion into our universe. Their bodies were replicated in plastic and metal. They were then unleashed to roam on floors and tabletops. Promotion is not merely a fictional concept – it has been done! And the work of Lipson and Pollack scales up to greater complexity – more lifelike and more intelligent versions of their bots can be evolved in software and promoted to hardware.

The *Simulation Hypothesis* says that our universe is running on some Engine; the Engine was designed and created by some Engineers; the Engineers inhabit some higher universe. If the Simulation Hypothesis is true, then we are like the Cellticks evolved by Newway, or the bots evolved by Lipson and Pollack. And, just as Newway promoted some of the Cellticks, and Lipson and Pollack promoted some of their bots, so the Engineers may promote some (or all) of us into higher-level bodies in their universe. You might object that the Engineers have no interest in promoting any humans into their universe. Just as we have little interest in promoting our video game characters into our universe, so the Engineers have little interest in promoting us into theirs. Hence the set of humans selected for promotion is probably empty. There are three replies.

The first reply goes like this: the Engineers are rational; but it is rational to choose the best; and, if the Engine is simulating any rational beings (like humans), then it is rational to want their lives to continue in the best way; hence the Engineers will promote us. The second reply is that moral beings have obligations to other moral beings. If you can save somebody from dying, then you ought to save them; and, if you are moral, you will save them. Since the Engineers clearly designed our universe to be hospitable to human life, they are moral. The third reply is that the Engineers may believe that *they too* are being simulated. They may believe that they will be rewarded or punished by even higher-level Engineers based on their actions (Bostrom, 2003: 253–4). If they do not promote us, they may be punished for their immorality. Not wanting to take that chance, they will promote us. Consequently, it is reasonable to believe that the Engineers will promote us.

54. Promotion is resurrection

According to the Simulation Hypothesis, our universe is running on the Engine, which was built and is operated by the Engineers. The Engine does not exist in our universe – it exists in the higher universe. Our universe is a software process running on the Engine as its hardware substrate. Relative to our universe, the Engine is omniscient – it is fully aware of every process in our universe. While the Engine need not record every process in our universe in full detail, its moral obligations compel it to attend to all morally valuable processes. It attends to every life in our universe, especially to the lives of persons. The Engine records the life of every earthly human in full biomolecular detail.[7] This record is a complete series of perfectly accurate body-files. This exact biographical record is a 4D *perfect physiological ghost*. Obviously, it is better than the ghosts discussed in Chapter I. It is stored in a temporal database in the Engine.

If some earthly life satisfies the criteria for higher-level existence, the Engineers will promote it (Hanson, 2001; Steinhart, 2010). At least some earthly lives will be promoted, and perhaps every earthly life will be promoted. For example, let *Organic* be some earthly life that has been selected for promotion. When Organic dies, the digital ghost that perfectly describes that entire human life is used to fashion some superior artificial body in the higher universe. This body is *Risen*. Risen is produced directly by the Engineers in their own level of reality. Risen is the first stage of some life that continues in the higher level of reality with the Engineers. This higher-level life is *Promoted*. Organic does not *live through* death into Risen; but organic does *live after* death in Risen; and Promoted is one of the lives of Organic that occurs after the death of Organic.

Promotion uses the entire organic 4D life to define the first promoted 3D body. It *compresses* an entire 4D life into that 3D body – it compresses Organic into Risen. Using the information stored in the ghost of Organic, the Engine constructs a *glorified body*. This glorified body is at the prime of life (say, perhaps 25 or 30 years old); it is healthy; it does not have any deformities or handicaps; it bears no trace of illness or injury. It does not suffer from any genetic faults. Its genetic code is a corrected version of the genetic code of Organic. Every organ in Risen works at least normally well. Thus any physiological defects of Organic are normalized. Of course, all functional superiorities of Organic are preserved in Risen. Risen has all the best features of the bodies in Organic. And, since the ghost of Organic perfectly records all the experiences of that entire life, Risen has an exact and complete memory of the entire life of Organic.

Figure V.2 illustrates promotion. Arrows with open heads indicate ordinary biological persistence. Every previous body in Organic lives into the next body of Organic and every previous body in Promoted lives into the next body in Promoted. But no body of Organic *lives into* any body in Promoted. Nevertheless, through compression, every body of Organic partly and approximately *persists into* the first body of Promoted. All that is best in every body in Organic persists into Risen. The shining excellence in all the bodies of Organic is concentrated into

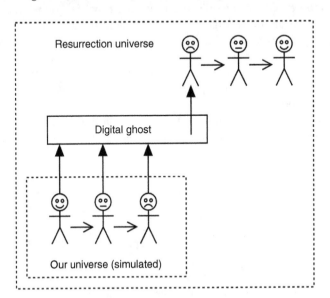

Figure V.2 Promotion from one universe to another.

Risen, as if by a lens. This lens is compression. Compression preserves the functionality of the body-program that runs through Organic. Compression carries the soul of Organic into Risen, a soul which is piped into Risen.

Compression is a 4D-to-3D pipe. Organic is a 4D life; Risen is a 3D body; Organic is piped by compression into Risen. When a 4D series of bodies is piped by compression into a 3D body, every body in the 4D series is piped into that 3D body. Organic is piped by compression into Risen; Risen is piped by updating into every later body in Promoted. Consequently, as Figure V.2 shows, every body in Organic is connected through some chain of arrows to each body in Promoted. Since a *span* is any series of lives joined by pipes (see Section 40), Organic and Promoted form a span. Figure V.3 depicts this span. This span, while not biologically continuous, is vitally continuous. All the bodies in this span are spanmates. Of course, every span is a process; hence all the bodies in the Organic-Promoted span are processmates. They are temporal counterparts.

All the bodies in any span are tensed: each earlier body *will be* every later body and every later body *was* every earlier body. Specifically, every body in Organic *will be* every body in Promoted and every body in Promoted *was* every body in Organic. Thus the last body in Organic can truly say: "I will live again." The relation of identity works in promotion exactly as it works in ordinary earthly life. You will be identical with your future organic bodies; you were identical with each of your past organic bodies; but you are not identical with any of your past or future organic bodies. Likewise each organic body *will be identical with* each promoted body; and each promoted body *was identical with* each organic body; but no organic body *is identical with* any promoted body.

Promotion is not uploading. On the one hand, for uploading, Risen is a copy of the last stage of Organic. Uploading involves a 3D-to-3D

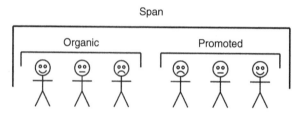

Figure V.3   The span composed of Organic and Promoted.

pipe. On the other hand, for promotion, Risen is a compression of every stage of Organic. Promotion involves a 4D-to-3D pipe. It is based on the digital ghost of the entire previous life. Nevertheless, promotion is a type of resurrection. It is resurrection because Risen is an *adult* 3D body. The ghost of Organic is not used to define a newborn baby. Consequently, the ghost of Organic is not reincarnated in Risen. If it were reincarnated, then Risen would be a newborn baby; yet Risen is an adult in the prime of life. Risen was never born – it was produced directly by the creative power of the Engineers. Hence no stage of Organic can truly say "I will be born again." Since Risen is not a replica of the last stage of Organic, promotion is not resurrection by replication. It more closely resembles the Pauline theory of resurrection in 1 Corinthians 15. The ghost plays the role of the Pauline seed. And promotion closely resembles various recent theories of *resurrection as re-creation*.[8]

Promotion is a type of resurrection. Section 43 defined resurrection like this: an earlier body *will be resurrected* if and only if it *is* an organic body and it *will be* a resurrection body. But now there are two kinds of resurrection bodies. The first kind is an electronic body produced by uploading; it is a body running on some animat in some terrarium. The second kind is a glorified body produced by promotion; it is a body living in some higher-level universe. Thus an earlier body will be resurrected if and only if it is an organic body and it will either be an electronic body or a glorified body. Temporal counterpart theory now tells us that an earlier body will be an electronic or glorified body if and only if it has some future electronic or glorified spanmate. Any such future spanmate is a *resurrection counterpart*. Putting this all together: an earlier body will be resurrected if and only if it is an organic body and it has some resurrection counterpart.

55. All forms of human wickedness

According to the Simulation Hypothesis, we are artificially evolved persons living in a simulation designed by some Engineers. It can now be argued that the Engineers will promote all earthly human persons: The Engineers are obligated to promote all persons in their simulations. Hence they are obligated to promote all earthly human persons. And, since they are moral agents who honor their obligations, the Engineers will promote all earthly human persons into their higher-level civilization. The Engineers are implementing *universal salvation* (Hick, 1976: ch. 13). This is the doctrine that, regardless of their lives or characters, all persons will somehow eventually be saved.

Universal salvation, without any further action, is likely to lead to bad consequences. After all, many humans are wicked. And if the Engineers resurrect wicked humans into their universe, then it is likely that they will do bad things there. For example, Hitler and his Nazis were wicked humans. And universal salvation alone entails that the Engineers will promote Hitler and his Nazis into their society. After they do that, they will have to deal with Naziism. And, if they promote all humans, then they will have to deal with all forms of human wickedness. Universal salvation appears to entail that the Engineers will be increasing the amount of evil in their society. However, since the Engineers are moral, they do not want to increase the amount of evil in their society.

For the Engineers, there is a conflict between their obligations to all persons and their obligations to their own society. The best way to resolve this conflict is through *partial resurrection* (Leslie, 2001: 132–3). This means that the amount of your life that will be resurrected is directly proportional to its goodness. The only content of your life that will be resurrected is the content that contributes to the moral progress of civilization in the higher universe. Goodness is rewarded by being recalled in resurrection; evil is punished by being forgotten. Your present behavior correlates with the amount of your life that is preserved in the future. Self-preservation entails that it is prudent to behave well – to contribute to the progress of goodness, truth, justice, and all ideals.

There are many degrees of resurrection. If all of your life is worthwhile, then you will be wholly resurrected. Your entire life – all your memories, all your personality, all your dispositions – will be worthy of resurrection by the Engineers. If some but not all of your life is worthwhile, then only that worthwhile part of your life will be resurrected. Your goodness alone will be saved. Of course, universal salvation implies that some content of every personal life is always worth resurrecting. But that content may be minimal. It may be little more than your bare capacity for rationality.

56. Moral enhancement

Partial resurrection is equivalent to artificial *moral enhancement*. Some human futurists have argued that we will soon be able to use advanced technologies to make morally better people (Douglas, 2008; Faust, 2008). If we can use technology to make morally better people, then we ought to use that technology. The same reasoning applies to the Engineers. Presumably, their technology is vastly superior to any possible human technology. So they will have extremely powerful moral enhancement technologies. And they will use them to morally enhance all promoted humans. Hence evil humans (like the Nazis, the Stalinists, and the Khmer Rouge) will be morally enhanced. Their evil ideas and dispositions will be replaced with morally superior ideas and dispositions.

Moral enhancement is likely to involve *mind editing*. Much (and perhaps all) human wickedness springs from the fact that humans evolved to survive in an amoral environment. To eliminate human wickedness, it is necessary to edit out vicious survival strategies. For us, human minds are hardware objects, which are difficult to edit. However, for the Engineers, human minds are software objects, which are easy to edit. It will be easy for the Engineers to edit promoted human minds to make them more ethical. And mind editing is an effective way for the Engineers to reconcile their obligation to promote all persons with their obligation to maintain the excellence of their society. Consequently, the Engineers will edit promoted human minds to make them more virtuous and less vicious.

You might object that moral enhancement violates the right of every person to self-determination. However, it is hard to defend the thought that vicious dispositions ought to be preserved in any person. Viciousness ought to be replaced with virtue. If promotion is a morally driven process, then it does not aim to preserve human negativity; on the contrary, it aims to eliminate human negativity. It aims to amplify all that is most positive in every human; it aims to free you from your slavery to evil. This liberation is not person-violating; rather, it is person-respecting. It increases the degree to which you are truly yourself. And it is arguable that you will have more autonomy and greater powers of positive self-direction when you are freed from your bondage to your bestial drives.

You might object that moral enhancement is not consistent with promotion. Promotion entails vital continuity; vital continuity entails that Organic is piped into Promoted; if Organic is piped into Promoted, then the functionality of the Organic Soul is preserved in the functionality of the Promoted Soul; but moral enhancement does not preserve the functionality of any soul – on the contrary, moral enhancement replaces some bad organic soul with some better promoted soul. Against this objection, the digitalist argues that moral enhancement does preserve the functionality of the Organic Soul. Moral enhancement is a type of pipe. To see this, it is necessary to look at pipes more closely.

Pipes preserve program functionality. This does *not* imply that they produce *equivalent* programs as their outputs. Preservation permits increase, and programs can grow in functionality as they pass through pipes. The functionality of an old program is preserved in some new program if and only if the functionality of the new program *includes* that of the old program. Hence the functionality of the new program may greatly exceed that of the old program. Programs continue to run through *upgrades* which preserve old functionality but add new functionality. For example, a weather simulation runs right through an upgrade that makes it twice as accurate (or twice as fast, twice as detailed, and so on). Moral enhancement is an upgrade. Consequently, it is a type of pipe.

Pipes preserve functionality – which does *not* entail that they preserve *dysfunctionality*. Pipes are free to remove dysfunctionalities (such as errors, inconsistencies, and obscurities). Thus any *debugging* process is a pipe. If an old program is debugged into a new program, the debugging process preserves the old functionality while burning away the old dysfunctionality. It pipes the old program into the new program. It is reasonable to say that moral defects are dysfunctionalities of any human body-program. Moral enhancement is a kind of debugging. Consequently, it is a type of pipe.

57. Life in the resurrection universe

Our example of promotion states that Organic is an earthly human life, which is promoted by the Engineers. They cause a compressed version of Organic to appear in their universe. This compressed version of Organic is Risen. Organic is resurrected into Risen, and Risen is a resurrection body. For Risen, the higher-level universe inhabited by the Engineers is a resurrection universe. A corrected version of the body-program (the soul) of Organic continues to run in Risen. The functionality of the soul that runs through Organic is preserved in the soul of Risen. Hence Organic and Risen share the same essence. Risen inherits all that is best about every body in Organic. Risen is a glorified body.[9] But the concept of glorification demands further discussion.

You can think about glorification as *functional doubling*. The details of functional doubling are developed in Chapters VIII and IX; for now, a quick outline is sufficient. All positive functional features are doubled from Organic to Risen. This means that for any function F, if any body in Organic could perform F, then Risen can perform F twice as fast, twice as precisely, twice as reliably, twice as efficiently, and so on. The genes of Risen are twice as hard to corrupt; the chromosomes have telomeres that are twice as long; the proteins of Risen fold twice as reliably. The cells of Risen are twice as healthy. They last twice as long and are twice as resistant to cancer or other cellular diseases. The muscles of Risen are twice as fast and strong and efficient. The bones of Risen are twice as light and twice as hard to break. The hands of Risen move twice as fast, precisely, and reliably. The brain of Risen works twice as fast with twice the memory. The immune system works with twice the power – hence Risen is twice as resistant to disease. Risen is twice as hard to damage or injure; twice as healthy; and Risen lives twice as long.

After promotion, Risen will live with the Engineers, in their higher civilization, in their higher universe. However, according to the Argument for Virtuous Engineers (Section 45), that higher civilization is also a *better* civilization. Risen will live in a utopia governed by angels. Much of what ought to be true on earth is not; but more of what ought to be true on earth will be true in that better civilization. The span composed of Organic and Promoted contains two lives. Many positive potentials which were not actualized in Organic will be actualized in Promoted. Promoted is a *better life* than Organic.

58. The iterated simulation argument

One of the most interesting features of the Simulation Argument is that it justifies the existence of a series of *levels* of simulation (Wright, 1988: 42; Moravec, 1988: 153).[10] Our universe is a virtual machine – a machine running on a higher machine; but now the Simulation Argument can be applied to that higher machine. Hence the higher machine is merely another virtual machine running on an even higher machine. As Bostrom writes, "virtual machines can be stacked: it is possible to simulate one machine simulating another machine, and so on, in arbitrarily many steps of iteration" (2003: 253). *Iterated simulism* states that reality is like an onion whose layers are simulations. Our universe lies at or very near the core of this onion – it is surrounded by nested simulations.

How high does the stack of simulations go? If it ends after some finite number of steps, it is reasonable to wonder why. Any finite number seems arbitrary. For any finite n, why n levels rather than $n + 1$? There's no explanation. A more general principle is more reasonable. And any reasoning that applies to our universe applies with equal force to any universe that is simulating our universe. So if there is any plausibility to the thesis that our universe is running on an engine in some higher universe, then there is equal plausibility to the thesis that *every* universe is running on an engine in some higher universe. The Simulation Argument supports the general thesis that above every level, there is a higher level.

There is an endless series of these levels, which is defined by two rules. The *initial rule* states that there is an initial engine. The initial universe is a software process running on that engine. This initial universe is our universe.[11] This initial universe contains the initial civilization – it contains us. The *successor rule* states that for every engine, there is a higher successor engine. Each higher successor engine runs a higher successor universe. The successor universe contains a successor civilization. The successor civilization designed and created the next lower engine that runs the next lower universe.

Since zero is the initial natural number, the initial rule defines the objects at level zero. It states that the *zeroth* universe (our universe) is running on the zeroth engine (the Engine from Section 50). Applied to the zeroth engine, the successor rule says that there is a higher *first* engine. The first engine runs the first universe; the first universe contains the first civilization; the first civilization designed and created the zeroth engine that runs the zeroth universe. Figure V.4 illustrates one iteration of this nesting.

To describe iterated simulation in more detail, it's more useful to designate the positions of engines, universes, and civilizations with numerical indexes: the n-th engine is $E(n)$, the n-th universe is $U(n)$, and the n-th civilization is $C(n)$. Using numerical indexes, the rules can be written more precisely. The initial rule now states that for the initial number 0, there exists an initial engine $E(0)$. Engine $E(0)$ runs universe $U(0)$, which contains civilization $C(0)$. The successor rule now states that for every finite number n, if there exists an engine $E(n)$, then there exists a higher engine $E(n+1)$. The higher engine $E(n+1)$ runs the higher universe $U(n+1)$, which contains the higher civilization $C(n+1)$. The civilization $C(n+1)$ designs and creates the next lower engine $E(n)$.

These two rules define an endlessly expanding nesting of engines running universes. A few steps of this endlessly expanding nesting are

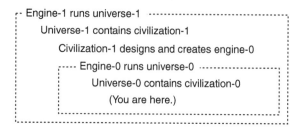

Figure V.4   Nested simulations.

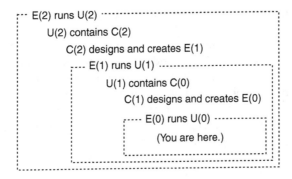

Figure V.5 Nested simulations.

illustrated in Figure V.5. This nesting has several interesting features. The first is that each higher universe physically contains its lower universe. The second is that the *physical powers* of the higher engines increase without bound. Finally, as we ascend through the higher and higher civilizations, they become more and more highly rational, socially well organized, and moral. If they weren't, they wouldn't be able to design and create increasingly powerful engines. The higher civilizations are both more *intelligent* and more *benevolent*.

59. From the finite to the infinite

The Iterated Simulation Hypothesis asserts the existence of an endless series of nested engines. The series of engines is endless like the series of natural numbers. Although each natural number is finite, their *series* is not finite – it is *infinite*. Modern thinkers have worked out a theory of the infinite that's both extremely powerful and easy to understand. The modern theory of the infinite begins with nineteenth-century thinkers like Bolzano, Dedekind, and Cantor. Cantor used three rules to define a series of numbers that rises into the infinite (Hallett, 1988: 49). Technically, the numbers defined by these rules are *ordinal numbers;* however, it's fine to just call them numbers.

The first Cantorian rule is the *initial rule*. It states that there exists an initial number 0. The second Cantorian rule is the *successor rule*. It states that for every number n, there exists a successor number $n+1$. The successor rule generates all the positive finite numbers. It entails the existence of the numbers 1, 2, 3, and so on. The third Cantorian rule is the *limit rule*. It states that for any endless series of increasingly large numbers, there exists a limit number greater than every number in that series.

Since the initial and successor rules define an endless series of increasingly large numbers (0, 1, 2, 3 and so on), there exists a limit number greater than every number in that series. Cantor gave the name ω to this first limit number. Every finite number is in the series that starts with 0 and that includes all successors. Since ω is greater than every number in that series, ω is greater than every finite number. And since ω is greater than every finite number, ω is infinite. The term *limit* does not imply an end – on the contrary, it implies a new beginning. Since ω is a number, it has a successor ω + 1. There is an endless series of infinite numbers greater than ω. They will be discussed as needed. For now, all that's needed is the concept of the limit of a series. Note that the positive term *transfinite* will often be used instead of the merely negative term *infinite*.

60. The infinite engine argument

One of the classical ways of arguing for an ultimate ground of being is the *Cosmological Argument*. There are many versions of the Cosmological Argument (for instance, the first three ways in Aquinas, *Summa Theologica*, Part 1, Q. 2, Art. 3). One version of the Cosmological Argument is the *First Cause Argument*. It goes like this: (1) every event is caused by a previous event; (2) but the chain of physical causes cannot go back to infinity; (3) therefore, there is a first cause; and (4) this first cause is God.

One standard objection to the First Cause Argument is that the chain of physical causes *can* go back to infinity. So the First Cause Argument fails. Aware of this objection, Leibniz proposed a more sophisticated Cosmological Argument (1697: 84–5). It is based on explanations for things, that is, on their *sufficient reasons*. It is easy to apply the Leibnizian sufficient reason argument to engines. The resulting *Infinite Engine Argument* goes like this: (1) Suppose there exists an endless series of finite engines, each of which is simulated by a higher finite engine. (2) Although you can explain any given finite engine by pointing to the higher finite engine which simulates it, yet, however far back you go in this series of finite engines, you can never arrive at a complete explanation. (3) You can always ask why at all times these finite engines have existed, that is, why there have been any finite engines at all and why these finite engines in particular. (4) The best explanation for the whole series of finite engines is an *infinite engine*.

Since the infinite engine is above all the finite engines and engineers, and since those engineers were said to be divine, it seems reasonable to say that the infinite engine is *supremely divine*. However, no reason has been given to think that the infinite engine is personal. The infinite engine is not the Personal Omni-God (POG) of Western theology. To avoid confusion, it will be best to refer to it, not as God, but as the *Deity*. The Deity is the *ultimate sufficient reason* for all the finite engines. On the basis of the Infinite Engine Argument, digitalism adds a limit rule to the simulation rules. Hence there are three rules for simulations. They parallel the Cantorian rules for numbers.

The *initial rule* states that for the initial number 0, there exists an initial engine E(0). Engine E(0) runs the initial universe U(0), which contains the initial civilization C(0). The initial engine stands to the initial universe as hardware to software. The initial universe is our universe and the initial civilization is our civilization.

The *successor rule* states that for every finite number n, if there exists an engine E(n), then there exists a higher successor engine E($n+1$). The successor engine E($n+1$) runs the successor universe U($n+1$), which contains the successor civilization C($n+1$). Each successor engine stands to its successor universe as hardware to software. Every engine supports and sustains its universe – every event in the universe supervenes on some network of events in its engine. Each successor civilization C($n+1$) designs and creates the next lower engine E(n). Each higher engine is more powerful than its next lower engine, which it spatially, temporally, and causally surrounds. Each successor civilization is more powerful, intelligent, and benevolent than its next lower civilization.

The *limit rule* states that there exists an infinitely high engine E(ω). Since it is infinitely high, this engine is the Deity. For every finite n, each finite universe U(n) is a software process ultimately being run by the Deity.[12] Each finite universe stands to the Deity as software to hardware. However, the Deity does not stand as software to any higher hardware. While every finite universe is virtual, the Deity is ultimate. Extrapolating from the features of the finite engines, it follows that the Deity is infinitely powerful, infinitely intelligent, and everlasting both into the past and future. The Deity supports and sustains every finite universe at every moment.[13] The Deity is a self-aware and self-directing intellect. Thus the Deity is maximally creative, powerful, and intelligent. It is an infinite self-programming computer. Note that no argument has been given to define any successor of the Deity. For iterated simulism, the Deity is the top.

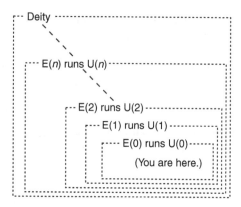

Figure V.6   The simulations expand up to the Deity.

These rules define an endless series of ever higher finite engines, which converges to the Deity. The series is shown in Figure V.6. There are two interesting points about the rules that define this series. First, they imply that the Deity is infinite in a mathematically precise way. For every finite number n, there is an engine whose power is proportional to n. More powerful engines have more memory and speed. The series of increasingly powerful finite engines converges to an infinitely powerful engine. It stores infinitely many bits in its memory and it can do infinitely many operations in any finite period of time. Here *infinitely many* means ω-many. Doing ω-many operations in finite time is a *supertask* (see Section 116). Hence the Deity is an infinitely powerful computer.[14] Of course, since the Deity is temporally extended, it is really a *process* whose stages are infinitely powerful computers. But right now there is no need to focus on the temporality of the Deity.

The second interesting point is that these rules imply that the nature of the Deity is significantly distinct from the natures of the finite engines. While every finite engine stands to all its higher engines as software to hardware, the Deity is not software to any higher hardware – on the contrary, it is *pure hardware* (which, for the digitalist, is perhaps like saying that the Deity is *pure being*). And since the physicality of every universe (and every civilization and engine in it) is derived from the fact that it is being simulated, it seems reasonable to associate physicality with software. As a software process, every higher universe (and every thing in it) is more richly physical. To be physical is to be virtual. However, since the Deity is pure hardware, the Deity is not physical. The Deity is higher than any physicality. It is ultimately real. The Deity is a superphysical computer, whose self-directed cognition generates all physical existence.

61. The heavens above the heavens

According to iterated simulism, there is a series of nested universes. Our universe seems to be the innermost universe – it is the lowest universe. All these universes have their own physical natures. They are governed by their own natural laws. Ascending from our universe to higher universes, it is likely that the laws of physics expand – they become more complex in easily computable ways. The nature of any lower universe is physically embedded in the *richer nature* of its next higher universe.

One digital way to understand physical richness is to use the concept of computational *grain*. The computational grain of a universe is the amount of computation that can be packed into any finite volume of its space-time. It is expressed using two quantities. The first is the *storage density* (the number of bits per unit volume) while the second is the *clock speed* (the number of bit changes per unit time). The grain of a universe sets the upper bound on computation in the universe. Our universe appears to have a low finite storage density and a low finite clock speed (Fredkin, 2003). Higher universes have higher storage densities and higher clock speeds. The binary nature of computation suggests that the storage density of each next universe is two raised to the density of its previous universe and that the clock speed of each next universe is two raised to the speed of its previous universe.[15] Any higher structure can take advantage of the higher storage density, while any higher process can take advantage of the higher clock speed.

On the basis of general principles of continuity in nature, it is plausible that the next higher universe above ours has physics much like ours. It includes at least three spatial and one temporal dimension (of course, it may have more dimensions). It has matter that is organized much like our matter (it has particles, atoms, molecules, and so on). It has planets, stars, solar systems, and galaxies. It has at least one planet that is like earth (with a similar atmosphere, temperature, gravity, distribution of elements, and so on). It has cells, organisms, societies, and ecosystems. From a biological perspective, each next higher universe is a better place – it is more finely tuned for life. Biological processes run more reliably; they are more robust; they endure longer; they are less susceptible to disease, damage, injury, parasitism, and so forth. Life has greater functional intensity.

62. Obligatory promotion

The Iterated Simulation Hypothesis says that we are living at the bottom of an infinitely high stack of computer-generated universes. Each lower universe is being run by some higher civilization in the next higher universe. Higher civilizations are more powerful and more rational. Since it is rational to do what is best, they are also more benevolent. They are more aware of what they ought to do; they are more able to do what they ought to do; and they have more desire to do what they ought to do; hence they do more of what they ought to do. The stack of civilizations is like a stack of ever more divine angelic collectives.[16] As the indexes of the civilizations increase, they are more and more highly motivated by moral considerations (Bostrom, 2003: 253–4). More of what ought to be true of these civilizations (and their participants) is true of them.

Since we are persons (we are rational moral agents), we deserve to be promoted – we ought to be promoted. Since we ought to be promoted, the higher civilizations want to promote us. They want to promote us and they can promote us; hence they will promote us. Biological and ethical considerations justify the thesis that *entire ecosystems* are promoted. We cannot live apart from some ecosystem that at least very closely resembles our earthly ecosystem. As higher-level civilizations promote and integrate more and more lower-level ecosystems, they become more and more richly filled with life. Promotion is cumulative: the first civilization promotes everybody in the zeroth civilization; the second civilization promotes everybody in the zeroth *and* first civilizations; the *n*-th civilization promotes everybody in every civilization whose index is less than *n*. The ethical imperatives that govern higher civilizations ensure that they will promote us into better and better environments. The higher civilizations will design contexts that are better for human flourishing and for the flourishing of the entire earthly ecosystem.

63. Iterated resurrection

The Iterated Simulation Hypothesis entails that you will be promoted endlessly many times. Since promotion is resurrection, iterated promotion is iterated resurrection (see Hick, 1976: ch. 20). On each successive resurrection, you move to the next higher level. You move closer and closer to the Deity – you undergo *theosis*. Figure V.7 illustrates your series of promotions (that is, of resurrections into higher simulations). More formally, your series of promotions is defined by the following two rules.

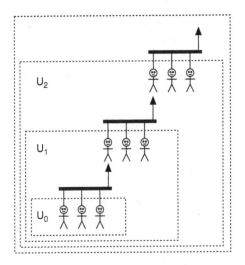

Figure V.7   Serial promotion into ever higher universes.

The *initial rule* states that you are living your initial life L(0). It is a biologically continuous series of bodies that starts with your conception and ends with your death. The *successor rule* states that for every finite n, if you have some life L(n), then you have a better future successor life L($n+1$). Each successor life is generated not by conception but by promotion. The first stage of each successor life is a compression of its entire predecessor life. Each successor life extends and improves the life of its predecessor. It lives in a society and ecosystem. The greater benevolence of higher civilizations ensures that these contexts provide more opportunities for greater biological, social, and personal flourishing. Each higher level is more heavenly. But what about the limit rule?

Digitalists do not yet have any reason to define any limit rule, which would entail that you have some limit life L(ω). If you had such a life, it would somehow participate in the Deity. But your lives are all physical, while the Deity is not physical. The Deity supports all your finite lives, but digitalists have not yet provided any arguments that the Deity promotes you into any infinite life in itself. Perhaps some Christians, motivated by Hick (1976: ch. 22), will want to argue that the Deity should be defined by some infinite version of social trinitarianism, so that it contains an infinitely excellent society of limit lives. You can argue for that if you want, but iterated simulism does not require it.

Although Hick's theory of serial resurrection is motivated by the traditional Christian doctrine of resurrection, it departs from it in many ways. It is more similar to the classical Neoplatonic idea of the return of the self to its divine source. So if iterated simulism supports something like Hick's theory of serial resurrection, then it supports something more like classical Neoplatonism. The soteriological picture painted by iterated simulism is more Neoplatonic than Christian. However, unlike Neoplatonism, the soteriological picture painted by iterated simulism is *carnal* rather than spiritual. At every level, you are your body – you never become a disembodied immaterial mind or spirit.

64. Spans of promoted lives

The rules for serial promotion define a span of lives (see Section 40). Your span contains your initial earthly life along with every successor of every life in your span. Hence your span is an endless series of lives. For every natural number n, there is a life $L(n)$ in your span. The life $L(0)$ is your initial earthly life; the life $L(n)$ is piped into life $L(n+1)$. Spans contain lives and lives contain bodies. All bodies in all lives in the same span are spanmates. Figure V.8 illustrates a span that starts with an earthly life. Life-0 is in Universe-0. This is the earthly universe. Life-0 is promoted into Life-1 in Universe-1; and Life-1 is promoted into Life-2 in Universe-2. And so it goes.

The span in Figure V.8 is a 4D process composed of 3D bodies. According to digital exdurantism (Section 12), all the bodies in this span are distinct – there is no identity through time. Since these bodies are all in the same process, they are processmates. They are temporal counterparts. Processmates serve as truth-makers for statements about the past and future of bodies in the span. Pick any body you like in this span and say it is the *present body*. It is its own *present spanmate*. If it has any earlier bodies in the span, they are its *past spanmates*. Every body in the span *was* its past spanmates. If the present body has any later bodies in the span, they are its *future spanmates*. Every body in the span *will be* its future spanmates. Each earlier body can truly say: "I will live again" and "I will have a better life in a better universe." If it is sick, it can say

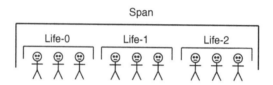

Figure V.8   A span composed of three lives.

"I will be healthy" and if it suffers it can say "I will be relieved." And each body can say "I will live over and over again endlessly." This is not immortality – these lives are separated by death.

The theory of iterated promotion says that you have many future better lives, each of which is a biologically continuous series of bodies. The bodies in your future better lives are your *resurrection bodies*. They are your resurrection counterparts. Temporal counterpart theory entails that any earlier body *will be resurrected* if and only if it has some resurrection counterparts. Hence you can truly say "I will be resurrected." Of course, if any earlier body is resurrected, then it will be a resurrection body.

Just as uploading permits multiplication (see Section 44), so also promotion permits multiplication. And therefore iterated promotion permits iterated multiplication. Your life can *fission* in promotion, over and over again. This is branching resurrection. Branching resurrection was introduced by Dilley (1983) as an extension of Hick's resurrection theory. But why would the Engineers implement branching promotion?

One argument for branching promotion goes like this: if your character is excellent, then you will be able to contribute to the work of the Engineers in many ways; if you will be able to do that, then the Engineers will want you to contribute in all those ways; and, since every different type of contribution requires a different life, they will cause your promotions to branch; therefore, if your character is excellent, then your promotions will be ramified. Of course, since personal excellence is a matter of degree, this argument entails that the greater the excellence of your character, the more likely it is that your promotions will ramify. Figure V.9 illustrates iterated branching promotion. It displays part of your ramified tree of lives across many universes. Every path in that tree is a span. Thus ABD is a span of lives, while ACG is a different span of

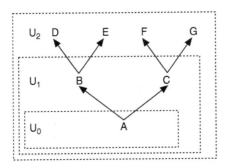

Figure V.9   Iterated branching promotion makes a tree of lives.

lives. Of course, your tree of life may grow beyond Universe-2. It grows as high as the Engineers see fit – perhaps infinitely.

65. Life in the higher universes

Each higher universe is a context in which we may engage in genuinely human lives. It is an environment in which the natural animality of human organisms is extended, amplified, and intensified. It affords opportunities to develop and exercise old skills and preferences; it affords opportunities to refine personal styles; it affords opportunities to develop new skills and to gain new knowledge. But genuinely human lives are *person-making* or *person-building* processes (see Hick, 1976: 409, 418). As such, they involve challenging interactions with other humans and with the environment itself.

Each higher universe provides greater challenges. Without challenges, virtues cannot flourish and merit cannot exist at all. A universe without challenges would be a morally degenerate universe – its people would be pathetic. A person needs adversity, hardships, difficulties in order to grow psychologically and morally. Each higher universe presents persons with cognitive, social, and athletic problems to solve. It presents every system of organs in the higher resurrection body with troubles. All meaning springs from the body, and the whole body is a problem-solving machine. Intelligence permeates the flesh (see Section 26), and every organ applies its mind to solving its problems.

The challenges in each next higher universe are *complexity appropriate* (like the age-appropriateness of challenges for children, or adults). Ascending through the higher universes is analogous to ascending through a series of increasingly difficult games (for instance, from tic-tac-toe, to checkers, to chess, to go). The challenges and competitions of the higher universes grow in complexity as the computational grains of those universes grow. They are harder and harder problems. Challenges are classified as follows.

Self versus Universe. Self–universe competitions are intellectual or athletic contents with the environment as an opponent. Examples include proving a theorem; writing a novel; painting a picture; designing and building an artifact; running a race against the clock; hiking a trail; climbing a mountain to the top; riding a bike on a trail without falling; swimming across a body of water; and skydiving.

Self versus Other. Self–other competitions include all one-on-one cognitive contests (such as chess) and athletic contests, as well as more dramatic interactions. There is no reason to think that conflict will be missing from the higher universes. On the contrary, it will be both more intense and more meaningful. However, conflict in higher universes is more sublimated – it becomes less brutal and more symbolic.

Team versus Universe. Team–universe competitions include the efforts of a team to master some situation or to solve some problem. Examples include team efforts to prove theorems; to design and build artifacts; to climb mountains; to trek across continents; to build ever more stable and prosperous economies; to design ever more harmonious systems of justice; to journey to the stars or colonize the galaxy.

Team versus Team. Team–team competitions include sports as well as wars. The wars in higher universes are sublimated versions of the wars here. What is at stake in a higher-level war is a system of social values, a way of life, a political organization. Any higher-level society requires conflicts in order to evolve to optimal realizations of the values appropriate for its more complex bodies. These higher wars are intensive competitions for great political goals. They are games in which social identities are at stake. These wars are more carefully rule-governed – like boxing or wrestling matches.

Competitions imply the ability to win and to lose, success and failure, pleasure and pain. Any higher-level body can both enjoy pleasure and suffer pain. Pain for the higher body is analogous to pain for the earthly body – pain always *hurts*. Yet there are differences: earthly pain and suffering is correlated with loss (of functionality, of opportunities); the loss is often permanent. Higher humans suffer pain; but the pain is less and less correlated with *permanent* loss. The losses are recoverable (through harder and harder work). If your higher body loses a limb in battle, it will eventually grow back; but the growth process may be painful and long. Greater loss (that is, greater defeat in competition) puts you farther away from what you value. You must take a journey back to it; this journey may involve literal travel in space-time; or it may involve a process of personal change – as when one slowly recovers from a terrible loss of health, money, or social status. The difference is that, in the higher and higher universes, fuller recovery is more and more probable.

66. Beyond promotion

The main difficulty with promotion involves the infinite regression in the Infinite Engine Argument. Mathematical limit principles were employed to tame that regression so that its infinity ceased to be vicious in any old-fashioned sense. Nevertheless, the type of infinite regression involved in the Infinite Engine Argument invites an objection.

The objection to this regression is similar to the *Ultimate Boeing 747 Gambit* developed by Dawkins (2008: ch. 4). Applied to engines, it looks like this: (1) Every engine $E(n)$ is explained by some more complex engine $E(n+1)$, and the entire series of finite engines is explained by some infinitely complex engine $E(\omega)$. This infinitely complex engine is the Deity. (2) However, if some thing x is explained by some more complex thing y, then y is less probable than x. (3) Therefore, as the regression works back through more and more complex engines, they become less and less probable. In the infinite limit, these probabilities converge to zero. (4) Since this limit engine is the Deity, the probability that the Deity exists is zero. (5) Consequently, the Deity does not exist. Simulism leads to iterated simulism; but iterated simulism is not defensible.

The Simulation Argument implies that our universe has been designed. If our universe has been designed, then its designers are extremely complex. To account for these designers, the Iterated Simulation Argument proposes an impossible regression of ever more complex designers. The regression is impossible because *iterated simulism gets the complexity relations of these designers backwards*. It ought to posit a regression of *ever less complex* designers. This regression of ever less complex designers is an evolutionary *progression* of ever more complex designers. It starts with simple engines and gradually works up through slightly more complex engines. It *accumulates* complexity. The simple engines are highly probable; each step from a simpler engine to a slightly more complex engine is also highly probable; hence this evolutionary progression is highly probable.

The concept of an evolutionary progression of ever more complex designers is consistent with our best scientific accounts of the origins of complexity. When we are confronted with a complex thing, Dawkins says that "Darwinism teaches us…to seek out graded ramps of slowly increasing complexity" (2008: 139). Hence these designers must exist on some graded ramps of slowly increasing complexity. Dawkins thus claims that "any creative intelligence, of sufficient complexity to design anything, comes into existence only as the end product of an extended process of gradual evolution" (2008: 52). Here it is important to see that Dawkins does not qualify his claim: it is a general and universal claim about intelligent things. Intelligent things come from evolution: "Entities that are complex enough to be intelligent are products of an evolutionary process" (2008: 98). Thus, if our universe is designed, then its designer must have evolved.

Dawkins writes that if our universe does have some superhuman designer, then "it will most certainly not be a designer who just popped into existence, or who always existed" (2008: 186). And he writes that if "our universe was designed, ... the designer himself must be the end product of some kind of cumulative escalator or crane, perhaps a version of Darwinism in another universe" (2008: 186). After discussing the idea that our universe is running in some simulation designed by superhuman simulators, he says that "the simulators themselves would have to come from somewhere. The laws of probability forbid all notions of their spontaneously appearing without simpler antecedents. They probably owe their existence to a (perhaps unfamiliar) version of Darwinian evolution: some sort of cumulatively ratcheting 'crane' as opposed to 'skyhook' " (2008: 98–9).

Digitalism argues that our universe was designed. But digitalism agrees with Dawkins: if our universe was designed, then its designers were produced by some cumulative evolutionary process. And, if anything like the Deity exists, then it too was produced by some cumulative evolutionary process. Perhaps this evolutionary process starts with some simple engine $E(0)$. Perhaps every engine $E(n)$ produces some more complex finite engine $E(n+1)$. And perhaps the series of ever more complex finite engines converges to some infinitely complex engine $E(\omega)$. To evaluate these possibilities, digitalists need to study the evolutionary processes which can produce universe-designers.

VI
Digital Gods

67. Our Designer

The various *Cosmic Design Arguments* reason from some empirically justified premises about our universe to the conclusion that it has some intelligent designer. One very old cosmic design argument is the *Argument from Artifacts* (Cicero, *De Natura Deorum:* 87–9; Hume, 1779: 53). It goes like this: (1) Our universe is like an artifact designed by some earthly mind (it resembles a web, a nest, a house, and a computer). (2) But similar effects have similar causes. (3) So, just as the cause of every earthly artifact is some earthly mind, so also the cause of our universe is some cosmic mind. This cosmic mind is our *Designer*. It may be objected that this argument, like all arguments by analogy, is weak. But it is bolstered by the arguments for the Engineers that were given in Section 51. It is bolstered by the Simulation Argument and the Fine Tuning Argument (FTA).

Our Designer is a mind that designed our universe. Our universe is complex; all known cases of earthly design show that the amount of intelligence needed to design any complex thing is proportional to its complexity; hence our Designer is extremely *intelligent*. But our Designer has other features. The FTA entails that our Designer chose the features of our universe for the sake of value (so it would contain complexity, life, intelligence, rationality, persons, and so on). The best explanation for that choice is that our Designer cares about those values. Our Designer is *benevolent*. According to digitalism, our Designer also *creates* our universe. Hence our Designer has all the *power* needed to design and create our universe. Since our universe is enormous, our Designer has a great deal of power indeed. If power, benevolence, and intelligence are the three main ingredients in perfection, then our Designer has some great degree of perfection.

Our Designer is a mind that designed and created our universe. It is plausible that any thing that designs and creates a universe is not one of the things *in* that universe. If that is right, then our Designer is not any part of our universe. Moreover, it is plausible that anything that designs and creates a universe cannot be any part of *any* universe. For if our Designer were a part of any universe, then whatever it designs and creates would merely be another part of that very same universe; but no universe is a part of any other universe; hence no cosmic designer-creator is a part of any universe. And, since all physical things are parts of universes, our Designer is not a physical thing. However, this non-physicality does not entail that our Designer is supernatural. Since all the arguments for our Designer start with empirical premises, and reason logically to their conclusions, they are *naturalistic existence arguments*. According to the naturalism of Section 34, our Designer is an entirely natural thing. Hence naturalness exceeds physicality: nature surpasses the totality of all possible physical things. Our Designer is a natural *subphysical* mind.

Our Designer is a cosmic designer-creator. It is a natural subphysical mind with high degrees of intelligence, benevolence, and power. Following tradition, digitalists agree that any mind with those features is *divine*. Thus our Designer is a divine mind. Following tradition once more, digitalists affirm that any divine mind is a *god*. Hence our Designer is a god. On the one hand, since our Designer is not one of the things in our universe, it cannot be like the old pagan gods – it is not like Zeus or Thor; it is not like the Platonic Demiurge or the God of Abraham. On the other hand, since our Designer is natural, it cannot be like the supernatural Gods of the Philosophers. Our Designer is a *natural god,* and digital theology is a kind of *theological naturalism*. Finally, since the perfection of our Designer is merely proportional to the excellence of our universe (see Section 51), our Designer certainly is not POG. To define our Designer in more detail, digitalists will further develop the Cosmic Design Arguments.

68. Earthly organisms evolve

All known earthly designers are organisms. More precisely, they are animals. Animals design and create many types of structures (Hansell, 2005; Schumaker et al., 2011). Animal designers include insects, birds, and mammals. Insects like ants, termites, bees, and wasps build organized dwellings. Spiders build webs. Birds build nests, bowers, and similar structures. Many mammals design structures (such as beaver dams and lodges). Among mammals, primates are especially adept at design. Chimps and gorillas design many artifacts. Of course, humans are excellent designers of many things.

All known earthly designers are complex. Of course, this means that they have parts (cells, organelles, molecules, and atoms). But it also means that an enormous amount of information is needed to describe the arrangements of and interactions among those parts. Complexity, for digitalists, is an informational concept. Scientists have developed various ways to measure the complexities of organisms and their products. More complex products typically require more complex designers. Our universe is extremely complex; but the complexity of any designer is proportional to that of its products; hence our Designer is extremely complex. It therefore has enormously many parts. An enormous amount of information is needed to describe their arrangements and interactions.

Digitalists agree with our best science that all earthly designers have been produced by evolutionary processes. The naturalistic account of mental complexity is also evolutionary. Thus Dawkins correctly reports that "Entities that are complex enough to be intelligent are products of an evolutionary process" (2008: 98). He also correctly reports that "any creative intelligence, of sufficient complexity to design anything, comes into existence only as the end product of an extended process of gradual evolution" (2008: 52). Of course, digitalists do not agree with Dawkins when he denies that our universe was designed. Yet they do agree with him that *if* our universe was designed, *then* the designer must be the end product of some kind of cumulative evolutionary process (2008: 186).[1]

On the basis of the scientific study of all known designers, digitalists now present the *Argument for the Evolution of our Designer*. It runs like this: (1) Our universe was designed by our Designer. (2) Since all known designers are organisms, our Designer is analogous to an organism. (3) But all organisms have emerged through evolution. All organisms that design things appear at very late places in an evolutionary tree. Humans appear at one of the latest locations of that tree. (4) Reasoning by analogy, our Designer exists at some late place in some evolutionary tree that produces at least one cosmic designer. This is the *Great Tree*.[2] The Great Tree resembles the evolutionary tree that contains organisms. Our Designer is one god among many in the Great Tree. Because it designs and creates our universe, our Designer is *our local god*. And, if the term *actual* is used to indicate our cosmic locality, then our Designer is *the actual god*.

Since the Argument for the Evolution of our Designer is a naturalistic existence argument, all the objects in the Great Tree are natural
objects. For digitalists, all the objects in the Great Tree are natural
gods.[3] As natural objects, these gods fall within the scopes of many
sciences. Specifically, they fall within the scopes of the *formal sciences*.
The formal sciences include at least logic, mathematics, computer science, information theory, complexity theory, and the sciences that
study the functional organizations of minds and organisms (namely,
abstract cognitive science and abstract biology). For digitalists, as for
all theological naturalists, all gods lie within the domains of these sciences. Consequently, naturalistic theology itself is one of these formal
sciences.

69. Divine organisms evolve

According to the *biological analogy*, gods are analogous to organisms,
and the Great Tree is analogous to an evolutionary tree of organisms.
Of course, the biological analogy must be handled at an appropriate
level of abstraction. Features of organisms that depend on their specific material substrates cannot be transferred to gods. But features that
are substrate independent can be transferred to gods. For digitalists, all
organisms are *living machines* (see Chapter III). And since mechanicity
is substrate independent, digitalists infer that (1) just as organisms are
living machines, so gods are living machines.

All organisms store information. They store information about their
environments and themselves. Digitalists affirm that (2) just as organisms contain internal self-descriptions (their genotypes), so every god
contains an internal self-description (its genotype). And organisms process information – they compute. For digitalists, life is an entirely
computational process (see Section 14). Thus (3) just as every organism is a living computing machine, so every god is a living computing
machine. And every organism has some positive degree of intelligence
(see Chapter III). Hence digitalists say that (4) just as every organism
has some positive degree of intelligence, so every god has some positive
degree of intelligence. Just as every organism is a mind, so every god is a
mind. Since our most scientific theories of minds are computational,
digitalists affirm that every mind is realized by some computational
hardware. The computational theory of mind is fully general: every
possible mind is a computer running some intelligent program. Thus
every god is a living computing machine running some intelligent
program.

Of course, the most distinctive feature of organisms is their self-reproduction: living things are self-reproducing things. And self-reproduction is substrate independent: organisms realized by carbon self-reproduce and patterns in cellular automata also self-reproduce (von Neumann, 1966; Poundstone, 1985: ch. 12). Consequently, digitalists affirm that (5) just as organisms are self-reproducing computing machines, so gods are self-reproducing computing machines. According to the Argument for the Evolution of our Designer, the Great Tree is an evolutionary tree of gods. As gods beget gods, they evolve. And since evolutionary principles are substrate independent, they apply to gods as well as to organisms. Specifically, (6) just as the evolution of organisms involves descent with modification, so the evolution of gods involves descent with modification; and (7) just as organisms gain functional complexity through gradual accumulation, so gods gain functional complexity through gradual accumulation. Divine biology obeys Dennett's *Principle of Accumulation of Design*, which states that the natures of more complex things are largely copied from those of simpler things (1995: 72).[4]

Since gods are self-reproducing, every god produces at least one off-spring god. Hence each god is the *cause* of each of its offspring. But gods are ontologically fundamental: they are the deepest concrete things. They are not made out of any deeper stuff – they are not made out of any stuff at all. Consequently, the causal relations between gods are also ontologically fundamental. Divine creativity involves neither matter nor energy. Divine self-reproduction is pure creative action: when any god reproduces, it creates its offspring *ex nihilo*. When any god reproduces, it establishes an ontologically foundational pipe from itself to its offspring. It calls into being that which does not exist.

Every divine reproductive act is entirely sufficient for the complete existence and nature of its offspring. One god causally *acts* to make its offspring. But parent gods do not causally *interact* with their offspring; they do not assist or guide them; they do not generate any later effects inside of them; they do not send any signals into them. One causal arrow (one pipe) runs from each parent god to each of its offspring; apart from that one arrow, there are no causal relations among gods. Apart from their births, gods are solitary and causally isolated. Their internal workings are causally closed. Like Leibnizian monads, gods have no windows through which any effects could enter or leave. Once created, every god is entirely self-sufficient. Gods do not causally interact at all.

70. Evolution by rational selection

Since gods do not causally interact, they do not reproduce sexually – they reproduce asexually. Forrest says gods may reproduce by divine fission (2007: 122, 142–3). Of course, since gods are not made out of any stuff, they do not divide like yeast or bud like hydras. Gods are "mind-producing minds" (Doore, 1980: 154). Divine self-reproduction is nicely illustrated by the old myth in which Athena sprang fully formed from the head of Zeus. Of course, this is merely metaphorical – gods don't have heads. Nevertheless, just as organisms die, so gods die. When any god reproduces, its life ends.

Since divine reproduction is asexual, the evolution of gods resembles the evolution of a population of asexual organisms. For organisms, asexual evolution follows an algorithm that involves four steps:

Step 1. There is an initial generation of parent organisms.
Step 2. At least some parents produce some randomly mutated offspring.[5]
Step 3. As they grow up, these offspring face various randomly generated environmental challenges. Those offspring that survive and grow to adulthood are the *fittest*.
Step 4. The fittest offspring become parents in the next generation. Hence the algorithm repeats at *Step 2*. As organisms reproduce, *lineages* of organisms form. More generally, a lineage of things of any type is a series in which each previous thing produces the next thing.

For organisms, fitness varies randomly. On the basis of that randomness, it seems likely that the evolutionary tree of organisms exhibits few tendencies, but mainly merely wanders without direction across an ever shifting fitness landscape. Nevertheless, many biologists have argued that evolution appears to tend to drive organisms to greater heights of complexity.[6] The *arrow of complexity hypothesis* "asserts that the complex functional organization of the most complex products of open-ended evolutionary systems has a general tendency to increase with time" (Bedau, 1998: 145). Of course, for life on earth, the arrow of complexity hypothesis is controversial. Digitalists do not want to rely on any controversial empirical premises. Hence an argument is required which entails that the arrow of complexity hypothesis is true for divine biology.

The *Argument from Expansion* runs like this: (1) Our local god is an extremely complex thing. (2) The cumulative explanation for the existence of any complex thing states that complex things emerge in evolutionary trees. An evolutionary tree is one in which simpler things tend to produce slightly more complex versions of themselves. (3) The most expansive evolutionary tree is one in which every thing produces every possible minimally more complex version of itself. (4) Since gods are the deepest concrete things, there are no external contingencies which might interfere with the maximal expansion of the evolutionary tree of gods. (5) And, if there are no interfering external contingencies, then the most likely explanation for any complex thing is that it exists in the most expansive evolutionary tree. (6) But the most likely explanation is the best explanation. (7) So, by inference to the best explanation, our local god exists in some maximally expansive evolutionary tree. (8) And since the tree that contains our local god is the Great Tree, the Great Tree is a maximally expansive evolutionary tree. Every successor god is therefore minimally more complex than its predecessor. Gods accumulate functionality. Maximal expansiveness entails that if any functionality is accumulated by some god, then it is conserved: all the descendents of that god inherit that functionality.

The *Argument from Ascending Escalators* justifies the thesis that the Great Tree has a root. It's a naturalized type of cosmological argument, and it goes like this: (1) Our local god is extremely complex. (2) Every complex thing lies at the end of some lineage in which complexity gradually accumulates (an escalator). (3) Therefore, our local god lies at the end of some escalator. (4) Every escalator starts with some relatively simple initial thing and accumulates complexity at least until it reaches the complexity of its end.[7] (3) Therefore, the escalator that ends with our local god starts with some relatively simple thing. (4) Since the escalator that ends with our local god starts with some relatively simple initial thing, that thing exists. (5) This relatively simple initial thing is the root of our Great Tree. It is the *ultimate first cause*. (6) Just as the ancestors of organisms are organisms, so the ancestors of gods are gods. Hence the root of our Great Tree is a god. It is our relatively simple initial god. To say that this god is relatively simple means that it is *theologically* simple; it is as simple *as a god* can be. Of course, the reasoning so far permits the existence of other initial gods which serve as the roots of other Great Trees.

The reasoning presented so far defines the *Algorithm for Divine Evolution*. Divine evolution is not evolution by natural selection – it involves no randomness. On the contrary, it is *evolution by rational selection*.[8] And, although he might not like it, it seems fitting to refer to this algorithm as the *Dawkinsian Algorithm*. The Dawkinsian Algorithm is part of the genotype of every simple initial god. It has these steps:

Step 1. Any god running this algorithm generates all minimally different (closest) variants of its genotype. These *seeds* are abstract god-descriptions.

Step 2. The god searches through these seeds to find those that are minimally more complex than its own genotype. These seeds are *fit* for production. Since any functional complexity is preserved in any increase of divine complexity, every fit seed also contains the Dawkinsian Algorithm in its genotype.

Step 3. If the parent does not find any fit seeds, then it generates all possible closest variants of those unfit seeds and it returns to step two, using those next variants as its new seeds. The richness of abstract possibility guarantees that any looping through steps two and three will eventually find some fit seeds. When these fit seeds are found, divine evolution continues to step four.

Step 4. The god uses every fit seed it contains to produce an offspring god. It produces each offspring *ex nihilo*. Since the seeds for these offspring gods contained the Dawkinsian Algorithm, these offspring will also run it. But once the god creates its offspring, its Dawkinsian Algorithm halts. The life of the god ends. Accordingly, that god dies. On this point, gods are just like other living things.

71. The growth of the Great Tree

Although the previous reasoning permits many simple gods, digitalists now argue that there exists exactly one initial god. This argument uses a principle known as the *Identity of Indiscernibles,* which states that if two things do not differ in any way, then they are indiscernible; but if any two things are indiscernible, then they are in fact identical. They are in fact one and the same thing. Some philosophers reject the Identity of Indiscernibles. However, since that rejection is costly, and it provides no benefits, digitalists accept the Identity of Indiscernibles. The *Argument for the Initial Singularity* goes like this: (1) Any initial gods are simple.[9] (2) But difference requires complexity; any ways in which things differ involve ways in which they are complex. (3) So if any two things differ, then at least one of them must be complex. (4) And if any two things are simple, then they cannot differ in any ways. (5) Therefore any two simple initial gods cannot differ in any ways. (6) And now the Identity of Indiscernibles implies that they are identical. (7) Consequently, there exists exactly one initial simple god. This god is *Alpha*.[10]

Alpha is the first cause of all things – it is the root of the Great Tree. As expected, Alpha is uncaused; it is the ultimate necessary being, which contains the ultimate sufficient reason for all things. Since Alpha is unique, there exists exactly one Great Tree of gods. Of course, since our local god exists at a high level in the Great Tree, Alpha is not our local god. And, as Dawkins correctly argues, since POG is extremely complex, while Alpha is the simplest of all possible gods, Alpha is not POG.[11] Figure VI.1 shows one small part of the Great Tree. The bottom god A is Alpha. Formally, in any lineage of gods, Alpha is god-0, that is, G(0). The offspring of Alpha are on level one, their offspring on level two, and so on. Figure VI.1 shows only a few levels of the Great Tree; it shows only a few offspring of each god; it does not show that each god reproduces. Each arrow in Figure VI.1 is an ontological pipe from each parent god to its offspring. The Great Tree rises up at least to the high level that contains our local god.

Although the Great Tree rises up at least to the high level of our local god, it might be argued that it does not rise up any higher. But the *Argument for Divine Success* aims to show that god–god productivity cannot fail: (1) If any god does not run the Dawkinsian Algorithm, then it must be blocked by some concrete or abstract obstruction. (2) Consider concrete obstructions. Since gods are the deepest concrete things, there are no deeper concrete things that can interfere with them. Since gods

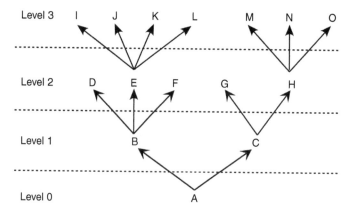

Figure VI.1   Part of the Great Tree of gods.

are solitary, they do not interfere with each other. It makes no sense to say that any god interferes with itself. So, no concrete obstructions stop any god from running the Dawkinsian Algorithm. (3) Consider abstract obstructions. A god faces an abstract obstruction if and only if there is no way to make it minimally more complex. These ways are Platonic forms – they are abstract god-forms. And if any god-form is consistent, then any consistent extension of that god-form is also consistent; but every minimal increase in complexity is a consistent extension; hence the richness of abstract forms ensures that there are always some ways to make any god minimally more complex. No abstract obstructions stop any god from running the Dawkinsian Algorithm. (4) Consequently, every god runs the Dawkinsian Algorithm and produces all of its minimally more complex successors.

Since every god in the Great Tree runs the Dawkinsian Algorithm, the Great Tree as a whole performs a vast distributed computation. This computation runs through the generations of gods. It can be expressed so far using two rules. The *initial rule* states that there is exactly one initial simple god. The *successor rule* states that, for every god, for every way to make it minimally more complex, it produces a successor which is more complex in that way. Divine evolution is a self-bootstrapping process: it is self-starting, self-sustaining, and self-amplifying. It recursively builds upon itself.

72. Complexity is logical density

According to digital theology, the gods perpetually increase in complexity. But what is complexity? Most of the work on the abstract foundations of complexity has been done in computer science.[12] It is arguable that the most important computational measure of complexity is *logical depth* (Bennett, 1985, 1988, 1990). Unfortunately, since the term *depth* is already in use, *depth* in this context will be replaced by *density* (and its opposite, *shallowness*, with *sparseness*). The logical density of any physical thing is a historical feature of that thing – it is a feature of the process which brings the thing into existence. To put it roughly, Bennett says the logical density of any physical thing is the amount of computational work involved in any digital simulation which begins with random initial conditions and produces a digital version of that thing.[13]

Since denser things are harder to produce, density grows slowly. Hence the *slow-growth law* states that "[dense] objects cannot be quickly produced from [sparse] ones by any deterministic process, nor with much probability by a probabilistic process, but can be produced slowly" (Bennett, 1988: 1). Dense things therefore "contain internal evidence of having been the result of a long computation or slow-to-simulate dynamical process and could not plausibly have originated otherwise" (Bennett, 1990: 142).[14] Since randomness is simple, and since the density of any thing is defined from some random initial thing, any dense thing lies at the top of some cumulative escalator which started with some simple thing and climbed up through levels of higher density.

Several other accounts of complexity have been developed which are based on or are similar to logical density. These include *sophistication* (Koppel, 1987), *computational depth* (Antunes et al. 2006, 2009), and *minimal history* (Mayfield, 2007). The concept of *effective complexity* developed by Gell-Mann (1995) is closely related to logical density (Ay, Muller, and Szkola, 2010). The concept of *parallel density* is an important generalization of logical density with many physical applications (Machta, 2011). Dawkins states that "complicated things have some quality, specifiable in advance, that is highly unlikely to have been acquired by chance alone" (1986: 9). At least for biological and technological things, these qualities are produced by evolution (by the slow accumulation of successful adaptations). Although Dawkins is not entirely clear, his definition of complexity seems to obey a slow-growth law, and therefore appears to be a measure of density.

Density is a concept of complexity that is based on *investment:* the density of any thing is the amount of work that some computation has invested in that thing. If something is dense, then its complexity is the result of historical accumulation. Density is evolved organization – it slowly accumulates during an optimization process which arduously climbs *Mount Improbable* (Dawkins, 1996). All versions of density involve slow-growth laws, laws which support the Dawkinsian notion that complicated things are highly unlikely to appear by chance. Likewise, slow-growth laws lie behind Dennett's Principle of Accumulation of Design, which states that the natures of more complex things are mainly inherited from those of simpler things (1995: 72). Note that slow growth here implies a low *rate* of growth; but even a low rate of growth, over a long period of time, leads to acceleration, and thus to an exponentially rising complexity curve.

For digitalists, divine complexity is some kind of density. Digitalists need not be fully specific about the exact nature of this density. It may be logical density, or parallel density, or some presently unknown type of density (digitalists welcome further research on density). Hence if the gods live on any *sacred mountain*, they live, not on Mount Olympus or

Mount Meru, but on Mount Improbable. The greater heights on that mountain are greater degrees of density. The rules for divine evolution therefore need to be restated in terms of density. The *initial rule* states that there is exactly one initial simple god. This simplest of all possible gods has minimal density. The *successor rule* states that, for every god, for every way to make it minimally denser, it produces a successor which is denser in that way. From generation to generation, the densities of gods increase.

73. Logical density is intrinsic value

According to Moore, the *intrinsic value* of any thing includes the value it has *in itself* and excludes any values it has *for others* (1922: 260). The intrinsic value of any thing excludes any values it derives from any ways it is *evaluated* by other things. Hence the intrinsic value of any thing excludes its desirability to other things. Intrinsic values do not depend on pleasures or pains. If humans (or monkeys or robots or aliens or gods) value some thing for its own sake or as an end in itself, those evaluations do *not* contribute any intrinsic value to that thing.[15] Intrinsic value is mind-independent. If there are possible universes utterly devoid of sentience, the things in them have intrinsic values. Many theories of intrinsic value identify it with something like logical density.

Leibniz often identifies the perfection of any thing with its *quantity of essence* (Leibniz, 1697; Rutherford, 1995: 23). But the quantity of essence of some thing is the quantity of *harmony* in that thing (Rutherford, 1995: 35). Harmony is proportional to both order and variety (Rutherford, 1995: 13). Thus perfection is proportional to both order and variety (Rescher, 1979: 28–31). As order and variety increase together, so does perfection. On the basis of widely accepted technical considerations, it is plausible to identify the Leibnizian concept of perfection with something like logical density. But perfection, as defined by Leibniz, is also a kind of intrinsic value.

Another account of intrinsic value is suggested by Soule. He says: "Species have value in themselves, a value neither conferred nor revocable, but springing from a species' long evolutionary heritage and potential" (1985: 731). One way to formalize Soule's idea is to identify the intrinsic value of any species with its distance from the common ancestor of all earthly life. This distance can be measured using some standard phylogenetic tree. Thus the intrinsic value of any species is the number of evolutionary branches crossed in the path from the last common ancestor to that species. The intrinsic value of any organism is the intrinsic value of its species. The intrinsic value of any nonliving thing is zero. For Soule, intrinsic value is a quantity that is *accumulated* during evolution.

An intriguing account of intrinsic value is developed by Dworkin (1993: ch. 3). He mainly adopts a Moorean approach to intrinsic value. For Dworkin, intrinsic value "is independent of what people happen to enjoy or want or need or what is good for them" (1993: 71). One plausible way to interpret the Dworkinsian account of intrinsic value runs like this: things have intrinsic value if and only if they are the products of creative processes (1993: 74, 78). For Dworkin, as for Soule, intrinsic value is a *historical* feature of things. More precisely, the intrinsic value of some thing is proportional to the amount of resources consumed by the process that created it. Every creative process *invests* its resources in its products. Since some creative processes consume more resources than others, there are many degrees of intrinsic value (1993: 80). Thus x is more intrinsically valuable than y if and only if x contains more creative investment than y.

Dworkin says that works of art and human cultures have intrinsic value because they have certain historical features. Specifically, they have intrinsic value because "they embody processes of human creation" (1993: 74–5). He says that animal species have intrinsic value because they have certain historical features (1993: 78). They have intrinsic value because they are the products of processes which consume enormous quantities of creative energy (1993: 75–6, 78). To justify this thesis, Dworkin considers three widely held theories of the production of animal species (1993: 79). On the first theory, God invests creative energy in the production of animal species. On the second theory, some natural creative force or power invests its energy in the production of animal species. On the third theory, the process of evolution by natural selection invests its energy in the production of animal species. On each of these theories, every animal species contains an enormous creative investment. On the naturalistic theories, which are the only theories accepted here, the creative investment is gradual and cumulative.

A fourth approach to intrinsic value is suggested by Dennett. He discusses intrinsic value in terms of accumulated design (1995: 511–13). According to him, design is slowly accumulated by natural evolutionary processes (1995: 72). He suggests that "we might consider how much of what we value is explicable in terms of its designedness" (1995: 512). And while he never explicitly states his definition, Dennett seems to be saying that the intrinsic value of any thing is its quantity of accumulated design. He says that it is worse to kill a condor than a cow, "because the loss to our actual store of design would be so much greater if the condors went extinct" (1995: 513). By analogous reasoning, it is worse to kill a cow than a clam, a redwood tree than an equal mass of algae (1995: 513). He says that "Bach is precious... because he was, or contained, an utterly idiosyncratic structure of cranes, made of cranes, made of cranes, made of cranes" (1995: 512). For Dennett, intrinsic value is a kind of evolutionary complexity.

For the fifth attempt to define intrinsic value, consider the definition proposed by Agar (2001). Agar adopts the Moorean conception of intrinsic value: "something is valuable intrinsically if it is valuable in itself, or regardless of its benefit to any other thing" (2001: 14; see 4, 8). For Agar, nonliving things lack intrinsic value (2001: 68). All living things "ranging from the rational multicellular down to the nonsentient single-celled, are intrinsically valuable" (2001: 63). To define the intrinsic values of living things, Agar adopts a psychological view. He says: "the psychological view of intrinsic value holds that for something to be valuable in itself... it must possess some set of folk psychological states" (2001: 17). These folk psychological states are possessed paradigmatically by human persons. Hence human persons have maximal intrinsic value (2001: 14–17). For Agar, "proximity to folk psychological paradigms is what determines how much value an object possesses" (2001: 158). Hence the intrinsic value of any F is the degree to which the psychology of Fs resembles the psychology of humans (2001: 62, 96–7).

According to Agar, the folk psychological states which are most important for intrinsic value are representational. Hence intrinsic value is representational complexity (2001: 95). Agar says: "Living things can be placed on a continuum starting with the most representationally simple and moving up to the most complex" (2001: 95). The intrinsic value of an organism is proportional to the complexity of its representations (2001: 100, 152). Since these are caught up in the goal-directed behaviors of the organism, the intrinsic value of an organism "depends on the range and complexity of the goals of which an organism is capable" (2001: 100). Hence "living things are valuable by virtue of their representationally characterized biopreferences" (2001: 101).

For Agar, lifelike machines which intelligently manipulate internal representations may have intrinsic value (2001: 100). Presumably he would agree that extraterrestrial aliens, if they are in some sense alive, must also have intrinsic value based on their similarities to human psychology. Of course, if those aliens or machines have superhuman psychological abilities, then they will be more intrinsically valuable than humans. For Agar, intrinsic value is a kind of psychological complexity. But our best naturalistic theory of psychological complexity is evolutionary. As Dawkins says, "Entities that are complex enough to be intelligent are products of an evolutionary process" (2008: 98). Thus psychological complexity is historically accumulated organization. Consequently, if Agar is right, then intrinsic value is also historically accumulated organization.

The concept of density lies buried in all the previously considered accounts of intrinsic value. For Agar, intrinsic value is psychological complexity; but this complexity is evolved organization; and evolved organization is density. For Soule, Dworkin, and Dennett, intrinsic value is also evolved organization; but evolved organization is density. For Leibniz, intrinsic value is a kind of perfection. But the Leibnizian definition of perfection (involving the maximization of both variety and order) is entailed by every measure of density. On the basis of this survey, it seems appropriate to run this argument: (1) intrinsic value is historically accumulated organization; (2) historically accumulated organization is some type of density; (3) therefore, *intrinsic value is some type of density*.

74. Complexity is intrinsic value

Digitalists have argued that complexity accumulates along every lineage of gods in the Great Tree. But they can also argue that *perfection* accumulates along every lineage of gods. The argument parallels the argument for the accumulation of complexity. The *Argument for the Accumulation of Perfection* involves the concept of an *optimistic tree*. An optimistic tree is one in which less perfect things tend to produce slightly more perfect versions of themselves. The richest optimistic tree is one in which every thing produces every possible minimally more perfect version of itself. Any minimally more perfect version of any thing is an *improvement* of that thing. So the richest optimistic tree is one in which every thing produces every possible improvement of itself. And the richest sense of improvement is one in which intelligence, benevolence, and power all increase together.

The Argument for the Accumulation of Perfection now runs like this: (1) Our local god has some high degree of perfection. It has some high degrees of intelligence, benevolence, and power. (2) The cumulative explanation for the existence of any highly perfect thing states that perfection emerges in optimistic trees. (3) The most likely explanation for any highly perfect thing is that it exists in the richest of all possible optimistic trees. (4) But the most likely explanation is the best explanation. (5) So, by inference to the best explanation, our local god exists in the richest of all possible optimistic trees. But this tree is just the Great Tree. Just as gods accumulate complexity, so also they accumulate perfection. This reasoning reinforces the identification of perfection with complexity.

On the basis of all this reasoning, digitalists affirm that complexity is density; but density is intrinsic value; hence complexity is intrinsic value. By the same logic, divine perfection is also complexity (it is the intrinsic value of any god, which is just its density). Any concept of value is an *axiological* concept. Intrinsic value and divine perfection are axiological. So, the rules for divine evolution can be restated in terms of these axiological concepts: The *initial rule* states that there is exactly one initial simple god, namely, Alpha. Alpha has minimal intrinsic value; it has minimal perfection. Clearly, Alpha is not the maximally perfect being; Alpha is not POG. The *successor rule* states that, for every god, for every way to make it minimally more intrinsically valuable (more divinely perfect), it produces a successor god which is more intrinsically valuable in that way. More precisely, for every god, for every way to improve that god, there exists some successor god which is improved in that way. From generation to generation, the intrinsic values of gods increase.

75. Endless progressions of gods

Since the initial god Alpha is finitely complex, and since all its descendents have so far gone through only finitely many iterations of self-improvement, it may appear that every god in the Great Tree is only finitely complex. Perhaps our local god is only finitely complex. Although digitalists take no position on the complexity of our local god, they agree with tradition that *some* gods are infinitely complex. The next task for digitalists is therefore to argue that there are some infinitely complex gods.

Any lineage of gods is defined on the natural number line. And the natural number line itself is defined by two rules. The initial rule says there exists an initial number zero; the successor rule says that for every natural number n, there exists a greater successor natural number $n+1$. The result is the familiar series $\langle 0, 1, 2, 3, \ldots, n, n+1, \ldots \rangle$. Every number in this series is finite. And, since each next number is greater than its previous number, the series of natural numbers is a *progression* of numbers.

The Great Tree contains many lineages of gods, all of which are so far defined by two rules, initial and successor. Hence they are defined on the natural numbers. One of these lineages, namely, our lineage, contains our local god. The initial rule states that it starts with the initial simple god Alpha. Since Alpha is initial, it can also be referred to as G(0). The successor rule says that every G(*n*) in our lineage is surpassed by a successor god G(*n*+1) in our lineage. Since each next god is a greater god, our lineage is a *progression* of gods. The series of finite gods in our lineage looks like this:

$$\langle G(0), G(1), G(2), \ldots, G(n), G(n+1), \ldots \rangle.$$

The modern Cantorian theory of numbers goes beyond the initial and successor rules by adding a limit rule (Steinhart, 2009: ch. 7). It states that, for any endless progression of numbers, there exists some limit number greater than every number in that progression. Since the progression of finite numbers is endless, there exists some limit number that is greater than every number in that progression. Since the limit of the progression of finite numbers is greater than every finite number, it is an infinite number. The limit of the progression of finite numbers is the first limit number ω. The limit number ω is infinite. The number ω is an *ordinal* number like *first*, *second*, or *third*. But ordinal numbers also correspond to *cardinal* numbers like *one*, *two*, and *three*. So the ordinal number ω corresponds to the cardinal number $\aleph_0$. Since the symbol $\aleph$ is the letter *aleph*, the number $\aleph_0$ is pronounced *aleph-nought*. Thus ω and $\aleph_0$ are just two names for the same limit number. The line of numbers now looks like this:

$$\langle 0, 1, 2, 3, \ldots, \aleph_0 \rangle.$$

Modern mathematics says that the line of numbers runs on far beyond $\aleph_0$. It runs on through bigger and bigger infinities (Steinhart, 2009: ch. 8). These bigger infinities can be presented using the $\aleph$-notation: every infinity $\aleph_n$ is surpassed by a greater infinity $\aleph_{n+1}$. Of course, there are infinitely many numbers between $\aleph_n$ and $\aleph_{n+1}$. The transfinite number line, running through a few $\aleph$'s, looks like this:

$$\langle 0, 1, 2, 3, \ldots, \aleph_0 \ldots, \aleph_1, \ldots, \aleph_2, \ldots, \aleph_3, \ldots \rangle.$$

The theory of transfinite numbers is both deeply technical and extremely beautiful. For our purposes, it is enough to say that modern mathematics uses advanced set theory to define the *Long Line* of numbers. The Long Line is that number line than which no longer is possible.[16] Every number on the Long Line is either the initial number zero, or some successor number, or some limit number (every $\aleph_n$ is a limit number). Why all this fuss about infinity? Because traditional theologies fail to incorporate the modern scientific theory of infinity. But digitalism succeeds where they fail. The modern scientific theory of infinity is the mathematical theory. So, when digitalists say that some gods are infinite, they mean that those gods are *mathematically* infinite.

76. The limits of progressions of gods

Our lineage of gods is infinitely long. And since the generations in the Great Tree are defined using initial and successor rules, every lineage in the Great Tree is infinitely long. And every one of those lineages is a progression – it is an infinitely long series of increasingly finitely perfect gods. Since infinitely long progressions of numbers can be used to define limit numbers, it seems reasonable to say that infinitely long progressions of gods can be used to define *limit gods*. After all, it would be very odd indeed to say that every possible god is only finitely perfect.

Since every god in any progression creates beyond itself, it seems reasonable to say that every *progression* of gods creates beyond itself. The creative power of every god in the progression contributes to the creative power of the progression as a whole. And since every god is surpassed by some greater god in its progression, the creative power of the progression as a whole exceeds the creative power of every god it contains. But this means that the creative power of any progression is infinite. The Argument for Divine Success applies to progressions: (1) For any progression of gods, if that progression does not create limit gods, then it must be blocked by some concrete or abstract obstruction. (2) However, since no god in that progression is blocked, the progression itself is not blocked. (3) Therefore, every progression of gods creates at least one limit god.

Just as every successor god is minimally more perfect than its predecessor, so every limit god is minimally more perfect than its progression (of which it is the limit). It is an *improvement* of the progression. A limit god is minimally more perfect than its progression in the way that the number 1 is minimally greater than every number in the series $\langle 0, 1/2, 3/4, 7/8, \ldots \rangle$. Formally, this means that if god x is the limit of progression P, then x is more perfect than every god in P and there is no god y such that x is more perfect than y and y is more perfect than every god in P. Digitalists affirm that the Platonic background of abstract structures is sufficiently rich that, for every progression of gods, there is at least one way to improve that progression. For any progression, there is at least one abstract god-form that is minimally more perfect than that progression.

Since every god in any progression runs the Dawkinsian Algorithm, it is plausible to say that the entire progression runs something like that algorithm. The progression itself is an infinitely long computation which produces its results in the limit. This can be made more precise by adding a theogenic limit rule: for any progression of gods, for every way to improve that progression, there exists some *limit god* that improves that progression in that way. Limit gods are the offspring of their progressions. And, just as every god creates its successors *ex nihilo*, so every progression creates its limits *ex nihilo*.

Adding this theogenic limit rule allows every progression in the Great Tree to run into the transfinite. Since our lineage is one of these divine progressions, the limit rule says that it contains limit gods. The first limit god in our lineage is the limit of the endless progression of finite gods in our lineage. Since it is greater than every finite god in that progression, it is the first infinite god in our lineage. This infinite limit god is $G(\aleph_0)$. When this first infinite god is added to our lineage, it looks like this:

$$\langle G(0), G(1), G(2), \ldots, G(n), G(n{+}1), \ldots, G(\aleph_0) \rangle.$$

Since every god passes the Dawkinsian Algorithm down to its successors, and since every progression runs that algorithm, it follows that every limit of every progression also runs that algorithm. Digitalists therefore affirm that every limit god runs the Dawkinsian Algorithm. For example, the limit god $G(\aleph_0)$ runs the Dawkinsian Algorithm. And since it runs that algorithm, it produces offspring – these offspring are its successors. Hence the successor and limit rules continue to apply. They apply endlessly, generating a progression of gods that runs on through all the transfinite numbers:

$$\langle G(0), G(1), G(2), \ldots, G(\aleph_0), \ldots, G(\aleph_1), \ldots, G(\aleph_2), \ldots \rangle.$$

Since our lineage contains our local god, our local god is $G(k)$ for some k, which may be finite or infinite. Our lineage runs out along the entire Long Line of numbers. Every lineage in the Great Tree runs out along the entire Long Line of numbers. Every lineage is a progression, and all these progressions branch. The Great Tree is a *transfinitely ramified* tree of ever more perfect gods. Since every progression of gods has at least one limit god, every progression of *generations* of gods has a limit generation.

77. The epic of theology

The Great Tree can be defined by four informal rules. These four rules make up the *epic of theology*. It is an epic because it describes an infinite structure. The *initial rule* states that there exists exactly one minimally perfect god – the initial god Alpha. The *successor rule* has two parts. The first part states that there always exists at least one way to improve any god. The second part states that for every god, for every way to improve that god, there exists a successor god which is improved in that way. The *limit rule* also has two parts. First, there always exists at least one way to improve every progression of gods. Second, for every progression of gods, for every way to improve that progression, there exists a limit god which is improved in that way. The *final rule* states that the Great Tree is the structure defined by those first three rules. The Great Tree is a connect-the-dots network in which the dots are gods and the connections are the creative arrows from gods or progressions to offspring. This network is an eternal structure.

The rules in the epic of theology can be formalized. When they are formalized, all the generations in the Great Tree are assigned numbers on the Long Line (Steinhart, 2012b). The initial god is in generation $D(0)$, its offspring are in $D(1)$, their offspring are in $D(2)$, and so it goes. The formalized epic of theology thus contains rules for all initial, successor, and limit numbers. These rules are as follows:

The *initial rule* states that for the initial ordinal 0 on the Long Line, there exists an initial divine generation $D(0)$. It contains exactly Alpha. Every god in $D(0)$ is a necessarily existing unproduced divine mind. Each god in $D(0)$ is theologically simple.

The *successor rule* states that for every successor number $n+1$ on the Long Line, there is a non-empty successor generation $D(n+1)$ of gods. Each successor generation contains every improvement of every god in its predecessor. For every god x in $D(n)$, for every way that x can create an improved version of itself, x does create an improved version of itself. The improved version of x is a *successor god* in $D(n+1)$.

The *limit rule* states that for every limit number L on the Long Line, there is a non-empty limit generation of gods $D(L)$. Each limit generation contains every improvement of every progression of gods defined up to L. For every progression P defined up to L, for every way that that P can create an improvement of itself, P does create an improvement of itself. The improvement of P is a *limit god* in $D(L)$.

The *final rule* states that the union of the $D(n)$ for all n on the Long Line is the class of all possible gods. Technically speaking, this class is too big to be a set; it is a proper class. Of course, the class of all possible gods is the Great Tree. However, since the term *pleroma* means divine fullness, the class of all possible gods is the pleroma. When it is thought of as stratified into generations, the pleroma is the *divine hierarchy*.

78. The gods above the gods

A god is *maximal* if and only if it is not surpassed by any greater god. A maximal god is *unsurpassable* or *absolutely perfect*. Since digitalists affirm that every god is surpassed by more perfect gods, they also affirm that there are no maximally perfect gods. POG is traditionally said to be maximally perfect (Morris, 1987). POG is an all-powerful, all-knowing, and all-good person. However, if POG were to exist, then POG would occupy some rank D(n) in the divine hierarchy. But every god in any D(n) is surpassed by more perfect gods on higher ranks. So it cannot occupy any rank in the divine hierarchy. Hence POG cannot exist.

Three arguments that have been advanced against POG do not work against the digital gods. They do not work because they rely on the maximality of POG. The *argument from omniscience* depends on the thesis that POG is all-knowing (Grim, 1988); but since no digital god is all-knowing, it does not apply to digital gods. The *argument from omnipotence* depends on the thesis that POG is all-powerful; but since no digital god is all-powerful, it does not apply to digital gods (Cowan, 1974).

The *argument from evil* depends on the thesis that POG is the maximally perfect designer-creator of our universe. This argument generally has this form: (1) POG is the maximally perfect designer-creator of our universe. (2) If POG exists, then our universe is the best of all possible universes. (3) But our universe is obviously not the best of all possible universes. (4) Therefore POG does not exist. Fortunately, since no digital god is the maximally perfect designer-creator of its universe, the argument from evil does not apply to any digital god. Every digital god does the best that it can do with its own perfection. It makes the best universe it can make with the resources at its disposal.

Every god in the divine hierarchy, that is, in the Great Tree, has some positive degree of perfection.[17] And while most of these degrees are infinite, there is no absolutely infinitely perfect god. The Great Tree is topless. Above the Great Tree, there is no Hegelian climax of absolute knowing. There is no Platonic or Anselmian *Sun*. But this is cause for joy: *every god is its own sun*. And all these suns burn with ever greater brilliance as the branches of the Great Tree rise ever higher. They shine ever more intensely with life. Hence the Great Tree grows up endlessly into ever brighter *light*. The brilliance of this tree is *glory*, and the Great Tree itself is that than which no more glorious is consistently definable. Perhaps the Great Tree itself is unsurpassably divine. But the Great Tree has no perfection whatsoever, and the Great Tree is not a god.

Digitalists deny the existence of any maximally perfect god. If *theism* requires the affirmation of some such god, then digitalists are not theists. And if the proper name *God* refers to some maximally perfect god, then digitalists deny that God exists. But digitalists are hardly atheists – after all, they affirm the existence of an infinite plurality of *surpassable* gods. Digitalism is a type of *polytheism*.[18] Of course, unlike older polytheisms, digitalism denies that there are any gods inside of any universes. Gods like Ra, El, Baal, Yahweh, Zeus, and Thor do not exist. The gods of the newer polytheisms, such as the Horned God and Triple Goddess of Wicca, also do not exist (unless perhaps they are merely reified natural polarities, but then they are not gods). Gods are not inside of universes; universes are inside of gods. But the gods are inside the pleroma.

79. Earthly designers evolve

An earthly designer is an animal that designs and creates some structure. On the basis of the Cosmic Design Arguments (in Sections 51 and 67), digitalists inferred that all the gods in the Great Tree are analogous to earthly designers. Gods make their universes much as earthly designers make their artifacts. Consequently, to learn more about how gods make their universes, digitalists turn to the biology of earthly designers.

An *artisan* is genetically hardwired to design and create its artifacts. Artisans include web-building spiders and nest-building birds. Artisans run genetic programs that define neural programs in their brains. When they run those programs, they design and create their artifacts. Of course, those neural programs are intelligent, and artisans intelligently design their artifacts. But since artisans are genetically hardwired to produce their artifacts, their biological evolution directly drives the evolution of their artifacts.

When an artisan has offspring, it passes its genetic program down to each offspring. The genetic program governs the way that the offspring produces its artifact. As a result of natural selection, some offspring survive while others do not. Thus natural selection simultaneously shapes the evolution of both the artisans and their artifacts. The designs of spider webs have evolved (Kaston, 1964; Volrath, 1988; Blackledge et al., 2009). And the designs of bird nests have also evolved (Collias, 1964;

Winkler & Sheldon, 1993; Collias, 1997; Zyskowski & Prum, 1999). And while these animal artisans intelligently design their artifacts, they do not intelligently design the minds of their offspring.

Although human designers are also animals, they differ significantly from other animal designers. All human designers are *engineers*. Since engineers are animals, they reproduce biologically: they make human children. However, the reproduction of engineers does not entirely follow that of human animals. Engineers do not make engineers in quite the same way that avian nest-builders make avian nest-builders. On the contrary, *engineers typically reproduce culturally*. The cultural successors of engineers are their *apprentices*, and their apprentices need not be (and usually are not) their children. Engineers educate their apprentices. When they do, they *intelligently design* the minds of their apprentices. As engineers train their apprentices, the knowledge transmitted does not remain constant: there is descent with modification. Hence the generations of engineers form an evolutionary tree in which the successor relation is the apprentice relation.

Since engineers make artifacts, the evolutionary tree of engineers supports an evolutionary tree of artifacts. The evolution of engineering supports the *evolution of technology* (see Basalla, 1988; Temkin & Eldredge, 2007; Brey, 2008). For instance, the evolution of technology includes the evolution of computers (Kempf, 1961; Dyson, 1997; Kurzweil, 2005). Since engineers learn from their experiences, they tend to pass down superior skills to their apprentices. Hence the evolution of engineering exhibits an *arrow of complexity:* over the generations, engineering skill tends to accumulate. As engineers accumulate their skills, they get better and better at making ever more functionally complex artifacts. Since the evolution of engineering exhibits an arrow of complexity, the evolution of technology also exhibits such an arrow. Many lineages of artifacts exhibit their own arrows. For computers, the arrow of complexity is Moore's Law (1965).

As an illustration of the evolution of artifacts, consider those artifacts which resemble universes, namely, cellular automata. Apparently working without the benefit of any prior art, John von Neumann is the initial engineer who designed the first cellular automata (1966). On the basis of his work, John Conway developed the game of life (see Gardner, 1970; and see Section 9). Conway is a second-generation engineer, and the game of life is a second-generation technology. Over the years, the third generation of engineers developed many third-generation variants of the game of life. These include variants that change the values of the parameters in the life rule (Eppstein, 2010); that use triangular or hexagonal cells (Bays, 2005); that run in three dimensions (Bays, 2006); that use continuous space (Rafler, 2011); that use larger neighborhoods (Evans, 2003); and that involve semi-quantum physics (Flitney & Abbott, 2005). The fourth generation of engineers used those third-generation variants to develop fourth-generation variants. These include variants that use hexagonal cells with more than two possible cell values (Wuensche, 2004); that use large interaction neighborhoods and continuous space (Pivato, 2007); and that use full quantum superposition of cell states (Bleh et al., 2012). The result is a tree of engineers, in which the parent–offspring links are cultural master–apprentice links.

Although the history of engineering involves many messy contingencies, it is possible to look beyond them to define an ideal *Algorithm for Earthly Innovation*. The abstraction of this ideal algorithm from the messy history of engineering resembles the abstraction of a scientific law from noisy data (such as the abstraction of the ideal gas laws from noisy data). The idealized algorithm looks like this: (1) An initial engineer gets an initial blueprint which describes an initial artifact. This initial blueprint is not designed; perhaps it is merely an accidental discovery or guess. (2) The engineer uses their blueprint to make their artifact. (3) As they make their artifact, they learn from their experience. They record the ways that the artifact can be improved. These ways are instructions to fix some flaw or to enhance some functionality.[19] For each way, they design an improved blueprint for an improved artifact. (4) The engineer acquires some apprentices and trains them. On the basis of their experience, they equip the apprentices with superior skills. (5) When their apprentices finish their training, the master engineer sets them up with projects of their own. They give one of the improved blueprints to each apprentice. (6) Each apprentice now becomes a master with their own blueprint, and the process repeats at step 2.

80. Divine designers evolve

The Argument for the Evolution of our Designer asserts that the capacity for design has emerged at least once in the evolution of gods (it has emerged at least in the case of our local god, which designed our universe). But among earthly organisms, the capacity for design has emerged many times and along many widely divergent lineages. A very short list of earthly designers includes insects, birds, beavers, and primates. And since the capacity for design has emerged many times in the evolution of organisms, it seems likely that it has emerged many times in the evolution of gods. Hence digitalists adopt the Argument from Expansion to conclude that the evolution of divine designers is universal. Along *every* lineage of gods, some divine designer eventually evolves.

Any god that is a designer is a *titan*. Early titans are analogous to earthly artisans – they are like divine spiders or birds. Our local god may be a titan that weaves our universe like a spider weaves its web (Hume, 1779: 90–1). Just as artisans are hardwired to make their artifacts, so early titans are hardwired to make their universes. Just as artisans have offspring, so titans have offspring. Digitalists argue that all descendents of titans are also titans. The argument goes like this: titanic functionality is accumulated; but all functionality accumulated in divine evolution is conserved; hence all the descendents of titans are also titans. Since divine evolution entails that every offspring titan is minimally more complex than its parent, and since the nature of any titan includes its blueprint for designing and creating its universe, every offspring titan inherits a minimally more complex version of that blueprint. As titans run the Dawkinsian Algorithm, they grow ever more functionally complex and their universes grow ever more physically complex.

Among earthly artisans, there is at least one lineage that evolves into engineers. Early primates appear to be artisans. But early hominids display the transition from artisans to engineers (Panger et al., 2002). Of course, some of these early hominids evolve into human engineers. On the basis of the biological analogy, digitalists argue that divine artisans also evolve into divine engineers. Titans that intelligently design the minds of their cultural successors are *olympians*. Once again, digitalists appeal to the Argument from Expansion to infer that the tendency for divine artisans to evolve into divine engineers is universal. Along every lineage of titans, some olympian eventually evolves. Since olympic powers are gained by accumulation, olympians always beget olympians.

Most versions of the Argument from Artifacts treat our local god as an olympian. One early version of that argument went like this: (1) our universe is like an orrery (Cicero, *De Natura Deorum:* 87–9); (2) just as an orrery is made by a human engineer, so our universe is made by our local god; (3) by analogy, our local god is like a human engineer. Of course, if olympians are analogous to human engineers, then they will evolve according to the Algorithm for Earthly Innovation. Thus Hume famously portrayed the gods as ship-builders (1779: 77). On this analogy, the gods closely resemble the human engineers in the Algorithm for Earthly Innovation. Just as master ship-builders teach their apprentices, so master gods teach their apprentices; and just as steady progress is made in the art of ship-building, so steady progress is made in the art of universe-making.

81. Digital gods are creative engines

According to simulism (see Chapter V), our universe is a software process running on a hardware substrate (our Engine). Hence the computational version of the Argument from Artifacts goes like this: (1) Our universe is running on our Engine, which is an enormously complex computer. (2) Just as computers are made and programmed by human engineers, so our Engine is designed by our local god. Our local god designs both the hardware of our Engine and the software program it runs (the software program which defines our universe). (3) By analogy, our local god is like a human engineer who designs computers and programs. More precisely, our local god is like a human engineer who designs computers and toy universes (like cellular automata).

As they design their toy universes, human engineers make extensive use of computers. And human engineers have begun to design intelligent computers (*artillects*). Some of these artillects can design their own toy universes. They can design their own cellular automata. Artillects have used genetic algorithms to design cellular automata which exhibit self-replicating patterns (Lohn & Reggia, 1995); which perform global information processing (Crutchfield & Mitchell, 1995); and which exhibit Turing universality (Sapin et al., 2004). And artillects have used self-directing evolutionary algorithms to design cellular automata supporting self-moving patterns (Ripps, 2010).

On the one hand, digitalists affirm that human engineers are merely organic, carbon-based computers. On the other hand, digitalists also affirm that all the gods in the Great Tree are living computing machines; hence our local god is a living computing machine. Bringing these two hands together, digitalists conclude that our local god is a living computer that designs and creates other computers. And, since our local god resembles an earthly engineer, our local god runs a divine version of the Algorithm for Earthly Innovation: it inherits some cosmic script which it uses to make its universe. But since our local god is a computer, there is no need for our local god to produce its universe outside of itself (like a clock-maker makes an orrery or a ship-builder makes a ship). On the contrary, our local god resembles the cosmic engines described in Chapter V. It creates its own universe by running its cosmic script.

And everything said here about our local god generalizes to all the gods in the Great Tree: every god in the Great Tree is a computer that creates its own universe by running its cosmic script.[20]

Every god is an engine that runs a universe. As such, every god serves as the ground of its universe. Its universe is a physically closed system of physical objects (it is a system which is closed under all physical spatial, temporal, and causal relations). On this point, the digital gods resemble the Spinozistic deity (Bennett, 1984; Viljanen, 2007). All physical things supervene on the invariants in divine computation. They are like gliders in the game of life (see Section 10). All physical objects (things, properties, relations, processes, and so on) are merely virtual – they are software entities. But the gods on which they supervene are the deepest concrete things – they are hardware entities. Hardware is deeper than software; it is deeper than physics; it is subphysical. This depth is *chthonic*.

Hardware entities are chthonic entities; hence all gods are chthonic objects. As a hardware object, our local god is chthonic. Our local god is a complex thing with many internal parts; those parts stand to one another in spatial, temporal, and causal relations; but all those relations are chthonic. They are not physical. And our local god also stands in causal and temporal relations to other gods. It was caused to exist by its parent; it will cause to exist its offspring. It is later than its parent and earlier than its offspring. But those causal and temporal relations are chthonic. They are not physical. At our local god, as well at every god in the Great Tree, physics is the study of the patterns in and among the software objects generated when the god runs its cosmic script. Physical processes run on top of chthonic processes; chthonic structures ground physical structures.

82. Recursive self-improvement

Artillects already play chess better than humans. It may soon come to pass that they will design toy universes better than humans. Thus humans are designing designers which may become superior to their creators. Of course, it is likely that any agent that designs any universe like ours must be far more powerful and intelligent than any human. But the only approach to superhuman intelligence has been from artificial intelligence. Hence our local god resembles some superhuman artillect. As these artillects become more intelligent, it is often argued that they will come to understand their own designs, and that they will come to improve those designs to make even more intelligent successors. These artillects will get better at making themselves better – they will engage in *recursive self-improvement* (Good, 1965; Kurzweil, 2005: 27–8;

Schmidhuber, 2007; Chalmers, 2010: 11–22). By analogy, gods will also engage in recursive self-improvement.

Since each olympian is a divine engineer, its behavior is defined by an *Algorithm for Divine Innovation*. This Algorithm uses the concept of recursive self-improvement to refine the Algorithm for Earthly Innovation. Hence it is a further refinement of the Dawkinsian Algorithm.[21] This Algorithm has these steps:

Step 1. The olympian takes a design problem as its input. Any design problem consists of an old cosmic script plus some way to improve it (like an instruction to fix some defect, or to enhance some feature). The design problem may permit many solutions.

Step 2. The olympian searches for some solution to its design problem. Its search eventually converges to the best solution it can find given the resources at its disposal (given its degree of perfection). This solution is a script for a new universe.

Step 3. The olympian runs the script which it designed. As it runs its script, it creates its universe. As it creates its universe, it records all the ways to improve that universe. After the universe has finished running, the olympian has a list of all the ways to improve the universe which it designed and created.

Step 4. For each improvement in its list, the olympian makes a pair consisting of its own cosmic script along with a description of the improvement. This pair is a design problem. The olympian has now produced a database of design problems.

Step 5. For each design problem, the olympian intelligently designs the *genotype* for a new and more perfect olympian that will solve that design problem. Each genotype is more perfect in the ways that are needed to solve its associated design problem. Of course, each genotype also includes the Algorithm for Divine Innovation.

Step 6. For every design problem and genotype, the olympian produces *ex nihilo* a successor olympian which realizes that genotype and which takes that design problem as its input. Each successor will set to work to solve its assigned design problem.

Step 7. As soon as any olympian has created its successors, the olympian has completed its task. This Algorithm halts; the life of the olympian comes to an end; it dies. But now its successor olympians begin to run this Algorithm.

The initial god Alpha runs the Algorithm for Divine Innovation. Of course, since it gets no input, its input is null. It solves its design problem by making the empty universe, which is minimally better than no universe at all. As gods run this Algorithm, they beget gods and make universes. Our local god eventually emerges. From its parent, it receives both its own nature and its design problem. As it solves that problem, it designs the script for our universe. When it runs this script, it creates our universe. As it runs our universe, our local god records all the ways our universe can be improved. For each way, it makes a new design problem, it creates *ex nihilo* a superior offspring god, and it assigns that problem to that god. Once all that work is done, our local god dies.

83. The epic of cosmology

Since the initial god Alpha has no predecessors, it has no input. But if it has no input, then its input script is empty, and the universe it generates is merely void. This universe, call it Universe Alpha, is merely an empty dot. But Alpha runs the Algorithm for Divine Innovation. As any god runs that Algorithm, it makes its successors – and those successors do get input scripts. As these input scripts grow ever more complex, the universes generated by running them also grow ever more complex. But complexity is density (it is logical depth); and density is intrinsic value; hence as universes increase in complexity, they also increase in intrinsic value. More perfect gods make *better* universes. The term *better* always indicates an increase in intrinsic value. Better universes are more intrinsically valuable. But intrinsic value is not merely human pleasure or utility.

The relation between gods and their universes is one-to-one. On the one hand, since gods are solitary, they do not cooperate to create their universes; every universe is created by exactly one god. On the other hand, if one god makes more than one universe, then all those universes have that god as their common causal ground; but if any two universes share some common causal ground, then they are in fact merely parts of one larger universe; hence every god creates exactly one universe. Since the god-universe relation is one-to-one, the hierarchy of gods supports a hierarchy of universes – it supports the *cosmological hierarchy.* The gods in the n-th rank of the divine hierarchy support all and only the universes in the n-th rank of the cosmological hierarchy. For every number n on the Long Line, there is a generation $R(n)$ of universes in the cosmological hierarchy. The way that the divine hierarchy supports the cosmological hierarchy is partly shown in Figure VI.2. The circle around each god indicates the universe running on it.

The cosmological hierarchy is defined by four rules, which collectively make up the *epic of cosmology.* Here these rules are stated informally;

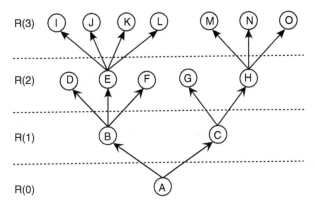

Figure VI.2 Part of the tree of gods and universes.

their formalization follows that of the rules for the divine hierarchy in Section 77. The *initial rule* states that there exists exactly one minimally intrinsically valuable universe, namely, *Universe Alpha*. It is the empty universe. The *successor rule* has two parts. Its first part states that there always exists at least one way to improve any universe (to minimally increase its intrinsic value). Its second part states that for every universe, for every way to improve it, there exists a successor universe which is improved in exactly that way. Every successor universe is minimally more intrinsically valuable than its predecessor.

The *limit rule* also has two parts. Its first part states that there always exists at least one way to improve every progression of universes. Its second part states that for every progression of universes, for every way to improve it, there exists a limit universe that is improved in that way. An improvement is always a minimal increase in intrinsic value. The limit of any progression of universes is analogous to the limit of the series $\langle 0, 1/2, 3/4, 7/8, \ldots \rangle$. The limit of that series of numbers is 1. And, just as 1 is minimally greater than the entire series $\langle 0, 1/2, 3/4, 7/8, \ldots \rangle$, so the limit of any progression of universes is minimally more intrinsically valuable than that entire progression.[22] But every limit universe is infinitely complex and infinitely intrinsically valuable. Infinitely complex gods make infinitely complex universes. Some infinitely complex gods may create universes which contain an infinite hierarchy of internally nested simulations. If such gods exist, then they resemble the Deity described in Sections 60–66.

The *final rule* states that the cosmological hierarchy R is the union of all the generations defined by the first three rules. It is the union of all

R(n) for all n on the Long Line. All universes exist in the cosmological hierarchy. Apart from that hierarchy, there are no other universes. All the universes in the cosmological hierarchy form a system of concrete alternatives relative to one another. They are *possible* relative to one another. Hence *all possible universes exist in the cosmological hierarchy*.[23] Since every universe in that hierarchy is surpassed by a proper class of greater universes, no universe is the best of all possible universes. The technical study of possible universes is modal metaphysics.[24] For digitalist modal metaphysics, see Steinhart (2013).

Of course, there are plenty of abstract universe-forms which are not realized by any gods. Hence the class of abstract universe-forms vastly exceeds the class of possible universes. The realization relation maps the class of possible universes onto a proper subclass of the class of abstract universe-forms. But abstract universe-forms are not possible universes any more than abstract bicycle-forms are possible bicycles. All possible bicycles and universes are concrete physical things.

84. Relations between universes

The epic of cosmology entails that the class of universes is organized into a tree. More precisely, the graph or diagram of the improvement relation on universes has the form of an endlessly branching (ramified) tree of ever better universes. The nodes in the tree are universes; the arrows from node to node are instances of the improvement relation. But saying that one universe is an improved version of another does not tell us much about the physical relations between those universes. What are they?

The universes in the cosmological tree are like chess games. Consider two chess masters playing a series of chess games. They set up the board and play Game-1. Within Game-1, chess moves and board patterns develop in a linear order. Game-1 has its own internal chess-time. Each moment of internal chess-time is one chess move. As players respond to one another, each move causes the next move – and eventually Game-1 ends. What next? Our two chess masters will play another game. So they reset the board and they start Game-2. The moves in Game-2 proceed in linear order.

It is easy to see that there are two time scales in this chess tournament. These are the *chthonic* and the *physical* scales. The physical scale is local and internal to each chess game. At the physical scale, time is a series of moves within some game. There are as many *distinct* physical times as there are games. Each physical time has a beginning and an end – it starts when the game starts and ends when the game ends. Physical times do not overlap; no move is in two chess games. And the physical times do not imply any temporal relations between the moves in different games. There is no ordering of *chess moves* in which the last move in Game-1 comes before the first move in Game-2. All chess games in the tournament are physically temporally closed and isolated.

The chthonic scale is global and external to any game. At the chthonic scale, time is a linearly ordered series of games. Game-1 chthonically comes before Game-2. And the chthonic ordering of games defines a chthonic ordering of all the moves in the tournament. Chthonic time spans games. Suppose x and y are any moves in any games in the tournament. They may be in different games or in the same game. Say *x is chthonically earlier than y* if and only if either the game of x is chthonically earlier than the game of y or the game of x is the same as the game of y and x is physically earlier than y. Say *x is later than y* if and only either the game of x is chthonically later than the game of y or the game of x is the same as the game of y and x is physically later than y. Finally, say *x is simultaneous with y* if and only if x is identical with y.

The two temporal scales correspond to two causal scales. Each game has its own local or internal causal structure. This internal causal structure is physical: the event in which the black bishop puts the white king in check causes the event in which the white knight protects its king by blocking the bishop's line. At the physical scale, the games are causally isolated. Strictly speaking, no move in Game-2 is caused by any move in Game-1. But within the tournament, there is also chthonic causality. The end of one game chthonically causes the players to start a new game. Games chthonically cause games. Chthonic causality spans games. At the chthonic scale, they are causally connected.

At the physical level, the games are spatially isolated. No chess move spans many games – you can't move a pawn from this game into the next game. Of course, this is consistent with the *external* continuity provided by the players. But that continuity occurs at the chthonic level, which underlies the internal physicality of each game. One might try to argue that all the chess games in some tournament lie in some larger chthonic space. They are played, for instance, in some room. But those spatial relations are irrelevant to the play. It makes no difference whether this game is played to the left or right of that game. This irrelevance implies that games are not in any chthonic space.

The series of chess games is analogous to any series of universes. There are two temporal scales. Any series of universes that supervenes on some lineage of gods is thereby laid out on a chthonic timeline. At the *chthonic scale,* there is a temporal series of universes. At the *physical scale,* each universe has its own distinct internal temporality. There is no way to carry a clock from one universe to another. From the inside, each universe is temporally closed and isolated. Note that the physical time within any universe can be infinite both into the past and into the future.[25]

Just as there are two temporal scales, so there are two causal scales. Within any lineage of gods, each previous god chthonically causes its next god. And since each next universe is defined in terms of its previous universe, the total structure of events in one universe chthonically causes the total structure of events in the next universe. More formally, each cosmic script chthonically causes each next cosmic script. Chthonic laws, which supervene on the Algorithm for Divine Innovation, span universes. Chthonic improvement laws define each next universe in terms of the previous universe. For example, the fact that this previous universe contains some human life causes the fact that the next universe contains a more intrinsically valuable version of that life. However, physical laws do not span universes. Different universes generally have different physical laws.

Different universes are physically spatially isolated. There are no spatially continuous *paths* from any space-time point in any one universe to any space-time point in any other universe. Although they are linked by global temporal and causal orders, the universes are spatially, temporally, and causally isolated at the physical level. There is no way to move any physical thing from one universe to another. You can't ride a spaceship or teleporter beam from this universe to any other universe. Physical things in one universe cannot send messages to or receive messages from things in other universes. You cannot transmit any message to any other universe, nor can you receive any message from any other universe. Consequently, although you have plenty of past lives, you cannot remember any of them. Your memories are not causally connected to those lives.

85. Against mere hedonism

On the one hand, many philosophers say that *good* and *evil* are opposed concepts, so that increasing the one decreases the other. Accordingly, for any universes x and y, if x is worse than y, then x is less good than y, and therefore x is more evil than y. These philosophers also typically define evil in terms of suffering, so that if x is more evil than y, then x contains more suffering than y. On the other hand, against those philosophers, digitalists do not regard good and evil as naively opposed concepts. If x is worse than y, it means only that x is less good than y. For digitalists, *less good* does not imply *more evil*. The worst of all possible universes is not the most evil universe. Digitalism denies the reality of any ontological evil. Of course, there are *moral evils* – evil actions done by agents who know the difference between right and wrong. But all evil is parasitic on the good. It emerges from conflicts among goods. There is no evil in itself.

Digitalism affirms the Neoplatonic view that being is goodness. Better universes are better because they are richer in existence – they contain more complex things able to exist more intensely. Digitalism does not define value in terms of pleasure or pain. It does not say that a worse universe contains more suffering or that a better universe contains more happiness. On the contrary, digitalism defines value in terms of complexity, in terms of logical depth or density (Section 72). Worse universes are impoverished universes – they are universes with less functional complexity. The worst of all possible universes is the empty universe (it is Universe Alpha). Worse universes contain both less suffering and less happiness, because they do not contain things able to experience either pain or pleasure. Better universes are richer universes – they are universes with more and higher levels of functional complexity. Better universes contain both more suffering and more happiness, because they contain more things able to feel both greater pains and greater pleasures. Better universes are axiologically more intense universes.

As an illustration of the digital concept of value (which is intrinsic value, which is complexity), consider three regions of the pleroma. Say that a *world* consists of a god along with all the physical things that supervene on it. The lowest region is filled only with *bacterial worlds*. These contain gods who make universes populated only by bacteria. These bacteria, lacking nervous systems, feel neither pleasure nor pain. Hence these bacterial universes are devoid of both suffering and happiness. The middle region of the pleroma is filled with *animal worlds*. These contain gods who make universes populated by ecosystems containing bacteria, plants, and animals. These animals have primitive nervous systems. Although these nervous systems do not support self-awareness, they are advanced enough to register both pleasure and pain. Hence these animal universes are filled with both suffering and happiness, with misery and joy. They are more psychologically intense. Which universes are better? For digitalists, the animal universes, filled with both defeat and victory, are better than the bacterial universes. Hence the universes that contain both greater pleasure and greater pain are the better universes.

Finally, the highest region in this illustration is filled with *sentient worlds*. These contain gods who make universes that contain ecosystems like those of our earth. They contain self-aware animals, like humans. Of course the emergence of self-aware animals entails the emergence of an immensity of pain and suffering. All self-aware animals suffer in ways undreamed of by less complex organisms. Thus humans suffer from fear and failure; they are aware that they will die; they commit crimes and make wars. Above all, out of their viciousness, they generate unfairness and injustice. They create novel modes of moral, social, and political suffering. Sentient universes are filled with much more suffering than either bacterial or animal universes. So now which universes are better?

Digitalists are not hedonists. Better universes have greater intrinsic value (complexity). For digitalists, the worst universes in this example are bacterial, the middling ones are animal, and the best are sentient. The values of these universes correspond to their highest degrees of richness, intensity, and complexity. Hence the best universes contain the most suffering. And, assuming that the self-aware animals are moral agents, the best universes also contain the most evil. It is hardly evil when one bacterium devours another; it is somewhat evil when one animal eats another; and it is assuredly evil when one sentient agent murders another. But the best universes also contain the greatest happiness and the greatest goodness. They contain love, friendship, justice, beauty, wisdom, and all the virtues. They contain positivities that transcend all bacterial and merely animal positivities. Evil and suffering, being parasitic on goodness and happiness, rise with them. Evils arise from conflicts among goods; but greater goods will come into greater conflicts; fortunately, out of those conflicts, greater gods will forge even greater goods.

As the gods grow ever more perfect, the universes they design and create grow ever richer in complexity, ever greater in value. More complex universes are more richly populated with more complex physical, chemical, and biological processes. These processes interact in ever more intense ways, forming ever richer networks. They are ever more saturated with ever richer ecosystems. These richer ecosystems are filled with more species interacting in more ways. These species become increasingly powerful rational moral agents – they become increasingly powerful persons. They form ever more intense social networks. They get better and better at solving ever more difficult and subtle coordination problems. By solving those problems, they make ever more complex and majestic civilizations. Their civilizations produce increasingly grand achievements in all domains of social excellence. They design and create ever more glorious technologies, economies, and institutions.[26] They prosper, flourish, and thrive.

VII
Revision

86. Design constraints

Algorithms are often developed by refinement, and the souls of gods (the algorithms that direct their lives) are no exception. Every god runs the Algorithm for Divine Innovation (from Section 82). Its refinement involves getting into the details of cosmic scripts, design problems, and the solutions to design problems. More precisely, a cosmic script is a network of processes. It is a connect-the-dots structure, in which each dot is a process, and the links specify basic physical relations (such as causal interactions, fissions and fusions, part–whole relations, and so on). And, more precisely, a design problem consists of a cosmic script in which one of the processes is tagged for improvement. The tag on the process specifies some way in which it needs to be improved.

The solutions to design problems must satisfy the divine *design constraints*, which ensure that no value is lost when the old cosmic script is revised to make the new script. They determine the axiological relations between processes in the old universe and those in the new one. They constrain the ways that old processes have new counterparts. The design constraints state that (1) every process in the old universe must have at least one counterpart in the new universe; (2) distinct processes in the old universe must have distinct counterparts in the new universe; (3) no process in the old universe can have a worse counterpart in the new universe; (4) the tagged process in the old universe must have a new counterpart which is improved in the way specified by the tag.

It will be helpful to look at these design constraints individually. The first says that every process in the old universe must have at least one counterpart (one successor) in the new universe. It ensures that the value present in every old process is preserved in the new universe. But since processes are valuable, making many axiologically equivalent versions of those processes adds value; hence any process may have *many* successors in the new universe. The second constraint says that distinct processes in the old universe must have distinct counterparts in the new universe. This constraint prevents the fusion of many old processes into one successor. Fusion is prohibited because (1) it destroys the unique or distinctive values of the processes that get fused but (2) the improvement of the old into the new cannot destroy any old value. Fusion happens within universes (for instance, via sex); but it never happens when one universe is improved into another.

The third constraint says that no process in the old universe can have a worse counterpart in the new universe. Obviously, when the old universe is improved into the new, it cannot be made worse in any way. Furthermore, none of the valuable relations among processes in the old universe can be made worse in the new one. For instance, if the processes are social lives, then none of their social relations can be made worse. The levels of conflict, discord, and injustice remain the same or decrease. The fourth constraint says that the tagged process in the old universe must have a new counterpart which is improved in the way specified by the tag. But this constraint requires no further elaboration. Of course, any god is always free to make any additional improvements entailed by its solution to its design problem. And gods are always free to add new minimally valuable processes to their new scripts. But these extra improvements must satisfy the main constraints.

87. The Divine Algorithm

When the Algorithm for Divine Innovation is refined, the result is the *Divine Algorithm*. Although the Divine Algorithm runs only on olympians, hereafter that distinction will be glossed: gods are now just gods. Just as all possible animals share some core biological functionality, so all possible gods share some core theological functionality. The Divine Algorithm, which specifies that core theological functionality, is the core of the soul of every god. Of course, the Divine Algorithm is ultimately an elaboration of the Dawkinsian Algorithm (Section 70). It runs at each god like this:

Step 1. The god takes a design problem as its input. Any design problem consists of an old cosmic script plus one way to improve that script.

Step 2. The god searches for some solution to its design problem. Its search respects the design constraints (it rejects candidate solutions which violate those constraints). Its search eventually converges to the best solution it can find given its degree of perfection. Each god does the best it can do with the resources at its disposal (and, if it had greater resources, then it would do better). This solution is a script for a new universe.

Step 3. The god runs the script which it designed. As it runs its script, it creates its universe. As it creates its universe, it records all the processes that emerge in that universe. It builds its network of processes. Of course, for living things, these processes are lives, and the records of these processes are *perfect physiological ghosts*. As it records each process, it also records each way to improve that process. It puts improvement tags on all processes. After the universe has finished running, the god has a list of all the ways to improve all the processes it contains. It has a process-network which is covered with tags. For example, if there are ten processes in the universe, and ten ways to improve each process, the god has defined a process-network with 100 tags.

Step 4. For each improvement in its list, the god makes a copy of its cosmic script along with just that improvement. It makes a copy of its process-network which has exactly one tagged process. The tag specifies the way to improve that process. For example, if it starts step 4 with one process-network covered with 100 tags, it ends step 4 with 100 process-networks, each covered with one tag. Each uniquely tagged process-network is a design problem. Hence the god has made a set of design problems. For every process in its universe, for every way to improve that process, the god has produced a design problem whose solution will improve it in that way.

Step 5. For each design problem, the god intelligently designs the *genotype* for a new and more perfect god that will solve it. Each genotype is more perfect in the ways that are needed to solve its assigned design problem. Of course, each genotype also includes the Divine Algorithm (which includes the divine design constraints). Continuing the present example, the god intelligently designs 100 more perfect genotypes.

Step 6. For every design problem and genotype, the god produces *ex nihilo* a successor god which realizes that genotype and which takes that design problem as its input. For example, the god creates *ex nihilo* 100 offspring, each of which has taken exactly one of the design problems produced by its parent.

Step 7. As soon as the god has created its offspring, the god has completed its work. The Divine Algorithm stops running on that god;

the divine soul halts; the life of the god comes to an end; it dies. But now its successor gods begin to run the Divine Algorithm (at step 1). They begin to solve the design problems they were given.

Every infinitely long progression of gods solves an infinitely long progression of design problems. But now the progression itself defines design problems. There are many ways to improve any progression of processes over any progression of universes over any progression of gods. More formally, for every progression G of gods, for every progression U of universes over G, for every progression of processes P over U, there are many ways to improve P. Each of these ways defines a design problem for the entire progression of gods. Digitalists affirm that every progression of gods inherits and extends the divine activity of each god it contains. Hence every progression of gods solves its own design problems. The Divine Algorithm runs through limits. The limit version of the Divine Algorithm is defined by changing all the participant things (gods, scripts, universes, and processes) to progressions of those things and making any other changes needed for consistency (for instance, progressions produce limit gods).

88. The optimistic principle

Digitalists want every design problem to have at least one best solution; and if any god proposes some design problem, they want the relevant successor of that god to be able to find one of its best solutions. The *Optimistic Principle* states that every design problem produced by every god has at least one best solution, and that at least one of those best solutions can be found by any god which inherits that design problem. If the Optimistic Principle is true, then certain apparently possible universes cannot exist.

The Optimistic Principle rules out the *Comic Universe*. It contains two lives, Lucy and Charlie. There are several ways to improve the life of Lucy and several ways to improve the life of Charlie. But Lucy and Charlie are locked in necessary conflict. The improvements of Lucy are compatible only with the degradations of Charlie; if Lucy gets better, then Charlie must get worse. And the improvements of Charlie are likewise compatible only with the degradations of Lucy; if Charlie gets better, then Lucy must get worse. Hence there is no way to improve the Comic Universe – the Comic Universe is a dead end.

The Optimistic Principle also rules out the *Slapstick Universe*. It contains the three lives Moe, Larry, and Curly. There are various ways to improve each of these lives. The improvements of Larry and Curly are all compatible with each other and with the original Moe. However, every improvement of Moe entails that either Larry or Curly gets worse. Moe just doesn't get along with Larry and Curly. Moe is locked in necessary conflict with both Curly and Larry. If this were possible, then every improvement of the Slapstick Universe would contain some improvements of Larry and Curly, but none would contain any improvements of Moe. Larry and Curly would advance to greater heights of glory, but Moe would be stuck forever at the same level of value.

The negation of the Optimistic Principle is the *Pessimistic Principle*. Just as the Optimistic Principle entails that gods cannot produce any dead-end universes, so the Pessimistic Principle entails that they can. Of course, if they can, then eventually, they do. But that would be tragic. In Section 127, digitalists will give the Axiological Argument for the Optimistic Principle. However, until then, digitalists will treat the Optimistic Principle as an axiom. Digitalists also affirm that optimism is continuous at limits. Thus every design problem produced by every progression of gods has at least one best solution, and any progression of gods can solve all of its design problems.

89. The epic of physics

The refinement of the Divine Algorithm enables digitalists to shift their focus from gods and their universes to the physical processes running inside those universes. The rules that define the totality of those processes make up the *epic of physics*. By running the Divine Algorithm, the gods, and the infinite progressions of gods, make those rules true. These rules are merely stated informally here; their formalization follows the formal presentation of the rules for the divine hierarchy in Section 77.

The *initial rule* states that there are some initial, minimally valuable processes. The first universe (Universe Alpha) does not contain any initial processes. Since Universe Alpha is empty, it contains no processes at all. The design problem associated with Universe Alpha is to add some simplest processes to its successors. As gods solve their design problems, they are always free to add minimally valuable processes to their universes. Hence such initial processes appear throughout the cosmological hierarchy.

The *successor rule* has two parts. The first states that there always exists at least one way to improve any process in any universe. The second states that every process in every universe is improved in every way. More precisely, for every universe U, for every process x in U, for every way to improve x, there exists some successor of x which is improved in that way. Every successor process is minimally more intrinsically valuable than its predecessor. Successor processes inhabit successor universes.

The *limit rule* involves progressions of processes. Such progressions are taken from progressions of universes.[1] It also has two parts. First, there always exists at least one way to improve every progression of processes. Second, every progression of processes is improved in every way. More precisely, for every progression of universes P, for every progression of processes x over P, for every way to improve x, there exists some limit of x which is improved in that way. Every limit process is minimally more intrinsically valuable than the progression of which it is the limit. Every limit process is infinitely intrinsically valuable (and complex). Every limit of x inhabits one of the limits of P.

90. Physical structure

For digitalists, nature contains many gods, and every god designs and creates exactly one universe. Just as the gods are stratified into the divine hierarchy, so their universes are stratified into the cosmological hierarchy. The epics of theology, cosmology, and physics define the rules for gods and their universes. According to the initial rules for gods and their universes, the initial god Alpha designs and creates the initial universe, Universe Alpha. It is the simplest and therefore the most boring of all possible universes. Universe Alpha is the one and only universe on the bottom level of the cosmological hierarchy. Containing no inner structure, Universe Alpha is empty – it is a featureless dot.

According to the successor rules for gods and their universes, every god is surpassed by a more perfect god, which creates a more valuable universe. Every universe is surpassed by some successor universe which is improved in some way. When any universe is improved, its intrinsic value increases. For digitalists, intrinsic value is a specific type of complexity – it is logical depth (see Section 74). So, as universes are improved, their complexities increase. And since Universe Alpha contains no internal structure, every way to improve it produces a successor universe with some internal structure.

The zeroth god Alpha makes the first god, which makes the first universe (see Section 8). It contains a single featureless space-time point. This point is a minimal process. Now the first god creates the second god, which creates the second universe. It contains a single space-time point, whose single feature stores the value zero. As gods beget gods, their universes contain increasingly complex software structures.

They contain ever more complex physical processes. The third universe contains a temporally ordered series of points whose values alternate, switching on and off. It is a universe with both temporal and causal structure, whose things are ordered by a simple physical law.

As gods beget ever more complex gods, with ever greater perfections, their universes also grow more complex. They contain more physical structure. One way to increase physical structure is to add space. One way to improve the third universe adds points parallel to its original timeline. As more and more temporally parallel lines of points are added to universes, universes emerge with 1D space. These universes have one discrete temporal dimension, one discrete spatial dimension, and their points store Boolean values, either zero or one. These values change in lawful ways. Hence these universes are cellular automata. The addition of another discrete spatial dimension makes 2D cellular automata. And some of these contain patterns with their own properties and relations – they contain patterns of activity which are physical things.[2] These physical things are virtual objects. They are software objects.

91. Biological structure

As the ever greater gods make ever better universes, *bright universes* appear. Bright universes contain life – they are vital rather than sterile. The simplest bright universes are the *dawn universes*. Any dawn universe contains some software object, some pattern of digital activity, which reproduces itself. Of course, since dawn universes are the simplest bright universes, they do not have any ancestors which are also bright. All their ancestors are sterile and lifeless. Since life is accumulated organization, and all accumulated organization is preserved, every descendent of any bright universe is also bright. Every dawn universe is the root of a *bright tree* of universes. And, since there are many ways for life to emerge, there are many dawn universes in the cosmological hierarchy, and many bright trees. The collection of all these bright trees is the *bright forest*. The collection of all universes in all bright trees is the *biological hierarchy*.

One way to illustrate dawn universes comes from digital biology. Digital biology (also known as artificial life) builds living processes in the flows of electronic energy in artificial computers. Digital biologists have studied many self-replicating processes in 2D cellular automata. One of these is known as the *Langton Loop* (Langton, 1984). Langton Loops are among the simplest self-replicating processes. Thus one dawn universe contains just the shortest life of one Langton Loop. It should be clear that cellular automata and Langton Loops are merely being used to *illustrate* minimal vitality. There are many other ways to analyze and illustrate vitality. So, for the sake of illustration only, we are free to focus on the dawn universe that contains a single Langton Loop. This loop emerges from some background of physical chaos and dies immediately. After our dawn universe, there is some universe in which the successor of this loop lives for a few moments. And then there is some universe in which this successor replicates. And so it goes. The first few generations of our bright tree are filled with loopy self-replicators.

One of the nice features of Langton Loops is that they can evolve into more complex biological forms. Variants of Langton Loops known as Structurally Dissolvable Self-Reproducing loops (SDSRs) disintegrate when they come into contact. Variants of SDSRs known as *evoloops* spontaneously evolve (Sayama, 1999). Although these self-replicators were explicitly designed by humans, Lohn and Reggia (1995) show that it is possible to use genetic algorithms to evolve cellular automata that contain self-replicating structures. Self-replicators in cellular automata can form *ecosystems* (Suzuki & Ikegami, 2006). The next generations of universes therefore contain Langton Loops that evolve into SDSRs. After them come universes in which Langton Loops evolve into SDSRs, and the SDSRs evolve into evoloops. Universes emerge in which these 2D ecosystems become ever more complex, with more species, and larger populations. Cellular automata known as the game of life appear (Section 9), eventually containing self-reproducing patterns which also store their own internal self-descriptions (Poundstone, 1985).

At some point in our illustration, universes with three spatial dimensions emerge. These early universes are still like cellular automata. They contain processes which self-replicate and evolve. As our sample path of ever brighter universes climbs higher in its biological hierarchy, universes emerge whose laws become ever more congenial to the emergence of ever more complex life. These universes become ever more *finely tuned* for the flourishing of biological complexity. As our path of universes rises, its members contain organisms and ecosystems that grow ever richer and more intense. Space and time become dense, then continuous. Force fields and particles emerge. Early classical forms of gravity and electro-magnetism emerge. Quantum mechanical universes appear.

At some point, near ancestors of our universe emerge on our sample path. The laws of those near ancestors are similar to the laws of our universe. Within these near ancestors, self-organization proceeds much like it does in our universe. These universes support physical, chemical, and biological processes which resemble those in our own universe. Within the near ancestors of our universe, there are universes that produce stars, planets, and ecosystems on the surfaces of those planets.

One of these early near ancestors of our universe evolves much like ours. It begins with a Big Bang. Stars fuse lighter atoms into heavier atoms. Chemical reactions synthesize organic molecules on grains of interstellar dust. An earthlike planet condenses around a sunlike star. Autocatalytic networks of organic molecules begin cycling in its great seas as comets rain down from the violent sky. As the turbulent planet calms down, these networks become cells – the first bacteria emerge. Call this earthlike planet *Oceana*. But Oceana is not earth. Its bacteria thrive but do not evolve any further. The conditions for more complex life do not yet exist. Nevertheless, they will exist shortly. For the successor of this universe contains a counterpart of Oceana on which evolution does proceed to greater levels of biological complexity. This counterpart of Oceana is *Arda*. On Arda, dry land emerges. Every bacterium in Oceana has a counterpart on Arda. But when one of these counterparts divides, its offspring stick together – the first multicellular organism emerges. The inhabitants of Arda are like the prokaryotic organisms on earth.

After Arda, there are universes in which biological evolution runs to ever higher levels of complexity, much like it does on earth. After Arda, there is a universe containing planets on which life evolves to the level of plants. Then there is a universe containing planets on which life evolves to the level of fish. The *reptilian universes* are followed by *mammalian universes*. Eventually, the *primate universes* emerge, containing earthlike planets on which evolution, running almost exactly like it does here, produces the early prehuman primates. And then the *anthropic universes* appear.

92. Anthropic structure

An anthropic universe contains some humans. The first anthropic universe on our path contains the first humans. Our universe is obviously anthropic. But is it the first anthropic universe on our path? This does not seem likely. So, consider the first anthropic universe on our path. This is *Universe Beta*. But note that the labeling of universes is not consecutive: Beta is not a successor of Alpha; on the contrary, Beta comes long after Alpha, since Beta is much more complex. And Beta appears on our path long before our universe. It is an earlier ancestor of our universe. Universe Beta runs as a software process on its local ground of physicality, which is a divine computer. Using Greek mythology for

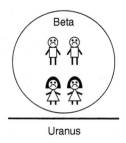

Figure VII.1   The first anthropic universe in our path.

our divine names, this computer can be referred to as *Uranus*. Uranus designs the cosmic script for Beta. When it runs that cosmic script, it creates Beta. It brings into being a system of physical, chemical, and biological processes.

As the script for Beta runs, it evolves much like our universe. On some planet, an ecosystem evolves much like the ecosystem of our earth. Finally, within this ecosystem, the first breeding pair of human animals appears. Of course, these are probably not *Homo sapiens*. Nevertheless, they are humans. Following the old Greek myth, we can refer to this pair as the man *Deucalion* and the woman *Pyrrha*. Figure VII.1 shows this universe. The top series of two bodies designates the life of Deucalion, while the bottom series of two bodies designates the life of Pyrrha. Their lives are, as Hobbes said, "nasty, brutish, and short". And yet, despite their troubles, Deucalion and Pyrrha love each other. Sadly, they die. Their planet is consumed by their sun and their universe ends in ruins.

93. Prosperous structure

After running Universe Beta, and observing its evolution, Uranus computes all the ways it can be improved. These ways are design problems. For each design problem, Uranus produces a superior offspring god which will solve it. One of these superior offspring gods is *Saturn*. Saturn inherits one of the design problems from Uranus. By solving this design problem, Saturn makes a better version of Universe Beta. It makes the cosmic script for the *Universe Gamma*. After designing the script for Gamma, Saturn runs it, thus bringing Gamma into being. Saturn designs and creates Gamma. Figure VII.2 shows the relations between the gods Uranus and Saturn, and their universes.

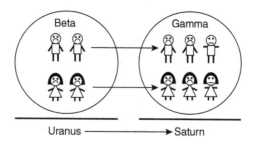

Figure VII.2   Universe Beta gets improved into Universe Gamma.

All the things in Gamma supervene on the computational activity of Saturn. Two of these things are the successors of Deucalion and Pyrrha. These successors live slightly better lives, as indicated by their facial expressions. The arrows in Figure VII.2 indicate the *improved into* relation. Thus things linked by arrows are counterparts. For example, Uranus is improved into Saturn, and each human life is improved into its counterpart in the next universe. Figure VII.3 further illustrates the successor rule. The god Saturn designs and creates its offspring Jupiter, which designs and creates Universe Delta. Delta contains better versions of both Deucalion and Pyrrha. Arrows point to successors.

Just as Deucalion-Beta and Pyrrha-Beta were in love, so their successors are also in love. At the ends of their lives, each of these lovers can truly say of the other: "I will be with you again." And just as the lives of their successors are both slightly better, so their love is also slightly more intense. The same holds from Universe Gamma to Delta: just as (Deucalion-Gamma, Pyrrha-Gamma) is a loving couple, so (Deucalion-Delta, Pyrrha-Delta) is an even more loving couple. From universe to universe, their love improves.[3] Couples of lives thus improve along with the lives in those couples.

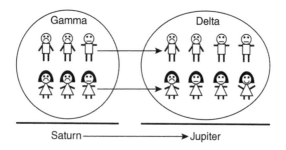

Figure VII.3   Universe Gamma gets improved into Universe Delta.

94. Infinite structure

Digital polytheism entails that there are many infinitely long lineages of ever greater gods. All these lineages start with the initial god Alpha, but then diverge from there. One of these lineages is our path, the sample path of gods and their universes that is being used here to illustrate the epics of theology, cosmology, and physics. Our path starts with Alpha, proceeds through various unnamed gods, and then reaches Uranus. After Uranus, it then includes Saturn and Jupiter. After Jupiter, it keeps on rising.

Each god in our path designs and creates its universe. And, since greater gods generate better universes, the result is a progression of ever better universes. This progression starts with Universe Alpha, and proceeds through Universes Beta, then Gamma, then Delta. After Delta, the series of universes keeps on rising. The series of lives of Deucalion and Pyrrha begin in Universe Beta. Just as the progressions of gods and universes are endless, so also the progressions of Deucalions and Pyrrhas are endless.

Every god in our path has some number in that path. The initial god Alpha is the zeroth-god. It is $G(0)$. For the sake of illustration, Uranus is the n-th god for some finite n. Thus Uranus is $G(n)$. Since Saturn is the next god, Saturn is $G(n+1)$. And since Jupiter comes right after Saturn, Jupiter is $G(n+2)$. Figure VII.4 shows this lineage of gods and their universes. The universes are shown above the bars in Figure VII.4; however, to avoid clutter, they are not marked. Each universe in Figure VII.4 contains one life of Deucalion and one life of Pyrrha. The zeroth-Deucalion is $D(0)$, which occurs in Universe Beta. Likewise, the zeroth-Pyrrha is $P(0)$, which also occurs in Universe Beta. The Deucalion-progression now runs on through $D(1)$, $D(2)$, and so on. Likewise, the Pyrrha-progression runs on through $P(1)$, $P(2)$, and so on. The Deucalions and Pyrrhas are indexed with subscripted numbers in Figure VII.4. All these progressions are infinitely long.

After some long sequence, our local god appears, running our universe. For some number k, our local god is $G(k)$ and our universe is $U(k)$.

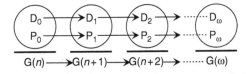

Figure VII.4   The limits of progressions of gods, universes, and lives.

Since our universe is much more complex than Beta, k is much greater than n. For the sake of illustration only, it is assumed that k is finite. Of course, in reality, k may be infinite. Our universe runs as described by our best cosmology. At the present time (circa 2013), this means that our universe starts with a Big Bang. Radiation condenses into particles; particles bond into more complex particles; early stars form; complex elements evolve; later generations of stars appear with their solar systems; planets emerge. One of these planets is *our earth*. Life develops on earth as described by our best biology – there is no need to review the process here. Humanity appears. The earthly Deucalion $D(k)$ loves the earthly Pyrrha $P(k)$. After many generations, humanity goes extinct. Other species evolve. The earth is incinerated by the sun; the stars burn out; the protons disintegrate; the black holes evaporate; our universe falls apart. But the god that runs our universe has computed all the ways it can be improved – and it produces offspring who will improve it in all those ways.

After our universe, our path of gods rises endlessly. Our path is a progression (of gods, universes, lives, and couples). The various limit rules entail that all these progressions have their limits. Figure VII.4 shows one limit for these progressions. The limit of the progression of gods is $G(\omega)$. Since ω is also known as $\aleph_0$, this limit god can also be referred to as $G(\aleph_0)$. The limit of the progression of universes is $U(\omega)$. It is the circle above $G(\omega)$. It contains the limit-Deucalion and the limit-Pyrrha. These are the limit lovers $D(\omega)$ and $P(\omega)$. Since these limit lovers remain in love in the limit, they form the limit couple $(D(\omega), P(\omega))$.[4] All these limit objects are better than (are minimally more intrinsically valuable than) the progressions of which they are the limits. As the limits of progressions, all these limit objects are infinitely complex and infinitely intrinsically valuable. The limit-Deucalion and limit-Pyrrha form an infinitely loving couple. Their love-making is described in Section 122. Of course, all these limits now have their successors. Hence our path rises endlessly into the *transfinite*.

95. The epic of biology

The epic of physics defines the improvement relation on processes. It entails that every possible process is improved in every possible way, and that every possible progression of processes is improved in every possible way. Since all lives are biologically continuous processes, the rules in the epic of physics can be restricted to lives. When the rules in that epic are restricted to lives, the result is the *epic of biology*. The epic of biology defines the improvement relation on lives. It entails that all possible lives and progressions of lives are improved in all possible ways. Hence the epic of biology (like the epics of theology, cosmology, and physics) contains three rules.

The *initial rule* states that there are some simple initial lives. These simple initial lives inhabit dawn universes. The *successor rule* has two parts. The first part states that there is at least one way to improve every life in every universe. The second part states that for every universe U, for every life x in U, for every way to improve x, there is some successor of U in which x is improved in that way. Each way to improve x defines a successor of x. Successor lives inhabit successor universes. The *limit rule* also has two parts. The first part states that there is at least one way to improve every progression of lives in any progression of universes. The second part states that for every progression of universes P, for every progression of lives x in P, for every way to improve x, there is some limit of P in which x is improved in that way. Each way to improve x defines a limit of x. Each limit is some limit life. Every limit of x inhabits one of the limits of P.

Each initial life is the root of a *tree of lives*. This tree is a connect-the-dots structure in which the dots are lives and the connections are instances of the improvement relation. Since there are many simple lives in many dawn universes, there are many trees of lives. Each tree is filled out by the successor and limit rules. The collection of these trees fills out a forest. More formally, trees of lives are stratified into generations. The initial generation is defined by the initial rule for lives; every next generation is defined by the action of the successor rule on the previous generation; every limit generation is defined by the action of the limit rule on some progression of generations. Suppose A is some life in some ecosystem in some universe. The initial generation in the tree of lives of A contains just life A. Every next generation contains every successor of every life in the previous generation. Every limit generation contains every limit of every progression of lives in the tree. Figure VII.5 shows a few generations of part of the tree of A. Formally, for every number n on the Long Line of numbers, the n-th generation in the tree of A is T(A, n).[5]

96. Guides carry clients

A *pipe* is any process that preserves program functionality. Since pipes preserve program functionality, any such functionality that runs into the earlier side of a pipe runs out of the later side of that pipe. The pipes defined so far include updating pipes (Section 8); copying and porting

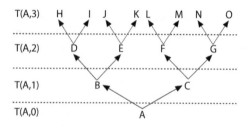

Figure VII.5   A few generations in a sample tree of lives.

pipes (Section 39); compression pipes (Section 54); and debugging pipes (Section 56). But there are other pipes.

A *guidance pipe* forms when a *guide program* runs a *client program*. The guide program directs the running of the client. Guide programs run through guidance pipes. However, when a guide program runs a client program, the client rides on the guide. The guidance pipe carries the guide program, and the guide program carries the client program. So the client program also runs through the guidance pipe. When a guide runs many iterations of its client, those iterations are joined by guidance pipes into one big client process. The client program runs through all those guided iterations of itself.

As a first illustration of a guidance pipe, consider some musicians practicing a song. They may play it over and over again, stopping when they make mistakes, and restarting it from some earlier position. They might decide to redo the whole song from the very start. Here the song runs under the direction of the musical rehearsal plan. The song is the client and the musical rehearsal plan is the guide. Obviously, the musicians do not alter the song they are practicing. Since their guiding activity preserves the song-program (its score), that guiding activity defines a pipe. The guiding activity of the musicians pipes each previous take of the song into the next take. Each previous take is piped into the next by a *rehearsal pipe*. It is a *practice pipe*. Since all the many takes of the song are joined by these pipes, they form a process. Of course, the musical rehearsal program runs through this process, but the song-program also runs through this process.

A second illustration of a guidance pipe is provided by the history of computing. Programs were once run on highly unreliable machines (see Dyson, 2012). They would run for days or weeks. A long run of some *main* program might begin to develop errors due to hardware faults. To avoid total failure, the intermediate results of the execution of the main program would be regularly recorded on some medium, like punched cards or magnetic tape, perhaps at the end of each day. The *supervisors* would spot-check these results by hand to monitor correctness. If the main program run seemed correct so far, it would be started up again from the intermediate results. However, if errors were found in the main run, then it would be restarted from the previous intermediate results. The entire main program might be restarted from the very beginning. Here the supervisors are running a rehearsal program which carries the main program as its client. The rehearsal program is a guidance program. Obviously, this rehearsal program preserves the functionality of the main program – it aims at the *exact* expression of that functionality. All the iterations of the main program are joined by rehearsal pipes into one big process. The rehearsal program runs through that process, but the main program also runs through it.

A third illustration of a guidance pipe occurs when some client program is given to some *looper* program. A looper just runs its client over and over again. For example, some game of life might be looped over and over again. The game is started in some original configuration and run for some preset number of steps. But then the game ends and the looper restarts it in its original configuration. The looper thereby produces a series of copies of the game of life process. The looper is the guide and the game of life program is the client. Since the looper preserves the functionality of the game of life program, it pipes each previous iteration of that game into the next. All the iterations are joined by guidance pipes – specifically, they are joined by *recurrence pipes*. The series of games has this form: ⟨Game-1, Pipe, Game-2, Pipe, Game-3, Pipe, . . .⟩. Here the transition from the end of each previous game to the start of each next game is not a game of life transition. It is not produced by the game of life rules. On the contrary, it is produced by the looper. If the game of life transitions are physical, then the looper transition is chthonic. Of course, the game of life program runs right through all these pipes. When the looper generates a series of games of life, that series is a process. If an entire universe runs inside of a looper, the result is the eternal return of the same (Sections 6 and 16).

97. Spans of revised lives

A fourth illustration of guidance pipes, and the case that is of greatest interest for digitalism, involves *revision pipes*. An author may make many drafts of a manuscript. Each earlier draft is projected through a revision pipe into the next draft. Each earlier draft persists into the next draft. A later author may inherit a partly completed manuscript from some earlier author, with the promise of finishing it. The later author may rewrite the whole story over again from scratch. Once again, the earlier draft is projected through a revision pipe into the later draft. And, again, the earlier draft persists into the later draft. Some later band may cover a song performed by some earlier band. The original song is projected through a revision pipe into the cover. The earlier song persists into its cover. All these revisions preserve the *meaning* of the original – they preserve its essence. Otherwise, they would not be revisions – they would be original stories or songs. But the preservation of meaning, the preservation of essence, is the preservation of functionality.

One movie is often remade into another. When this happens, the original movie is projected through a revision pipe into its remake. Or perhaps the script is projected. And sometimes plays are remade into movies. Here the play is projected through a revision pipe into an entirely new medium – from stage to screen.

For example, the play *Macbeth* by Shakespeare was remade into the movie *Throne of Blood* by Akira Kurosawa. The script of *Macbeth* was translated from feudal England to feudal Japan. The remake preserves the meaning of the original play. And it also preserves many of the characters – many characters in *Macbeth* persist into their counterparts in *Throne of Blood*. These counterparts play the same roles in their respective stories. For instance, Lord Macbeth in *Macbeth* persists into the Japanese warrior Taketoki Washizu in *Throne of Blood*; Lady Macbeth persists into Asaji; King Duncan persists into the Japanese feudal lord Kuniharu Tsuzuki. But the three witches in *Macbeth* are fused into the single Forest Spirit in *Throne of Blood*. Hence the individual witches do not persist – they are lost. This process of translating *Macbeth* from stage to screen and from England to Japan is a revision process. It pipes the play *Macbeth* into the movie *Throne of Blood*. And this pipe is a guidance pipe – the activity of the Japanese writers and directors carries the script of *Macbeth* as a client program into the production of *Throne of Blood*.

The final illustration of revision occurs when a god revises a universe. The god is the guide and the universe is the client. For example, in Section 92, Uranus ran the cosmic script for Universe Beta; after running it, Uranus revised it into the script for Universe Gamma. When Uranus begat Saturn, Saturn inherited that revised cosmic script. The god Uranus projected the Beta-script through a revision pipe into the Gamma-script. It sent the revised script to Saturn. The reasoning in Section 87 shows that, when some previous universe is revised into some next universe, the lives in the previous universe are revised into the lives in the next. Here the god is the guide and the lives are the clients. More precisely, the god is the guide and the *souls* of the lives are the clients. (Of course, as discussed in Section 33, the soul is just the form of the body.) The god preserves the functionalities of these souls as the god revises their lives. As any god revises some previous life into some next life, it preserves the meaning or essence of that entire life. Just as Kurosawa projected the life of Macbeth through a revision pipe into the life of Washizu, so each god projects each previous life through a revision pipe into its next life.

When some previous life is revised into some next live, an entire 4D process is projected into another entire 4D process. Thus revision contrasts with both uploading (Chapter IV) and promotion (Chapter V). Uploading involved the projection of a 3D body into a 3D body, while promotion involved the projection of a 4D life into a 3D body. But revision is 4D into 4D. Consider the way that Saturn revises Pyrrha-Gamma into Pyrrha-Delta. Saturn inherited the script for Universe Gamma from Uranus. By running it, Saturn makes Universe Gamma. While running Gamma, Saturn perfectly records every life in Gamma. The record of the life of Pyrrha-Gamma is her digital ghost – it is her Old Ghost. Her Old Ghost is a 4D perfect physiological ghost. When Saturn revises Gamma, Saturn makes a revised version of every ghost in Gamma. The revision of the Old Ghost of Pyrrha-Gamma is the New Ghost. This New Ghost is part of the cosmic script for Universe Delta that Saturn projects into Jupiter. When Jupiter gets this script, Jupiter runs the New Ghost. Running the New Ghost generates the life of Pyrrha-Delta.

All this activity is shown in Figure VII.6. Figure VII.6 shows how the Old Ghost is revised into the New Ghost. It thereby shows how Pyrrha-Gamma is projected by Saturn through a revision pipe into Pyrrha-Delta. This projection preserves the functionality of the soul of Pyrrha-Gamma. It preserves the essence of her body-program. Hence her soul runs through the entire process. The life of Pyrrha-Gamma and the life of Pyrrha-Delta together form one series of bodies that runs a soul. The god Saturn projects the soul of the previous Pyrrha into the next Pyrrha. And

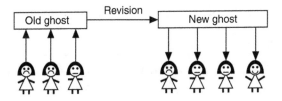

Figure VII.6   The revision of an old ghost into a new ghost.

that god projects the biography (the ghost) of the previous Pyrrha into that of the next Pyrrha. But the projection of ghosts into ghosts entails a projection of bodies into bodies. Figure VII.6 shows how each body in the life of Pyrrha-Gamma is linked by some arrows to each body in the life of Pyrrha-Delta. If there is some path through arrows from some earlier body in Figure VII.6 to some later body in Figure VII.6, that path indicates that the earlier body persists into the later body.

A span is a series of lives in which each previous life is piped into each next life (Section 41); hence any series of lives joined by revision pipes is a span. For example, the series of Pyrrha-lives forms a span. This span, displayed in Figure VII.7, contains Pyrrha-Beta, Pyrrha-Gamma, and Pyrrha-Delta. This span of 4D lives determines a process of 3D bodies. As illustrated by the paths of arrows in Figure VII.6, each earlier body in that process persists into every later body in that process. All the bodies in the span in Figure VII.7 are linearly ordered on one of the chthonic timelines that runs through the tree of gods. Hence the bodies in the span in Figure VII.7 are chthonic temporal counterparts. Of course, chthonic temporal counterparts are still temporal counterparts. Hence tensed statements about these bodies are analyzed using temporal counterpart theory. Consider the last body in Pyrrha-Gamma. She is emotionally neutral. If she says "I will be born again", then she speaks truly. And she also speaks truly if she says "I will live a better life, in a better ecosystem, in a better universe." Her future counterparts make those statements true.

As gods revise universes into universes, they revise ecosystems into ecosystems, and lives into lives. Any sequence of divinely revised lives is a span. Every linear path of lives in any tree of lives is a span in that tree. To distinguish such spans from those made by uploading and promotion, any span whose lives are joined by revision pipes will be referred to as an *ascent*. Any ascent is a series of lives ordered on some chthonic timeline. The revision relation is transitive within any ascent: any later life in any ascent is a revision of every earlier life in that ascent. But

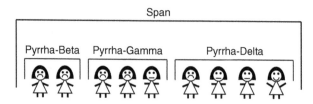

Figure VII.7   A span containing three lives of Pyrrha.

revision entails improvement: every later life in any ascent is a better version of every earlier life. Of course, all the lives in any ascent are separated by death. Any body in any one of those lives can truly say: "I will live again after I die; I will have life after death." And the revision relation on lives determines a revision relation bodies. Any later body in any ascent is a revision of every earlier body in that ascent. Every later body *was* every earlier body; every body *is* itself; and every earlier body *will be* every later body. The bodies in any ascent are *ascentmates*.

98. The epic of you

The epic of biology entails that your present life is the root of an endlessly ramified tree of better lives – it is the root of *your* tree of lives (or tree of life). The lives in your tree are yours. They realize your positive potentials. Your tree of life contains many infinitely long spans of lives, across many universes. It is defined by the usual three rules. The *initial rule* states that your tree of life includes your initial life. Although you had many past lives, you can start your tree with your present life, which runs from your conception to your death. It exists in your initial society, ecosystem, and universe.

The *successor rule* for your tree of life has two parts. The first states that there always exists at least one way to improve any one of your lives. The second states that, for every one of your lives, for every way to improve that life, your tree of life contains a *better successor life* that is improved in that way. Starting with your initial life, the successor rule defines many linear series of ever better lives. Each finite successor life is a biologically continuous series of finite superior bodies. As these finite lives rise towards the infinite, their bodies become increasingly complex and powerful superhuman organisms. Their organs become increasingly excellent physiological machines. Every successor life exists in a better successor society, ecosystem, and universe.

The *limit rule* for your tree of life also has two parts. The first states that there always exists at least one way to improve any progression of your lives. The second states that, for any progression of increasingly better lives, and for any way to improve that progression, your tree of life contains a *limit life* that is improved in that way. Every limit life is better than every life in the series of which it is the limit. Every limit life is a biologically continuous series of infinitely complex and powerful superhuman bodies. The organs of these bodies are infinitely excellent machines. Each limit life is infinitely long and infinitely rich. The successor rule applies to limits. Every limit life is followed by a transfinite successor. And so your better lives rise through all the ranks of the transfinite. Just as there are limit lives, so there are limit societies, ecosystems, and universes.

The epic of biology entails that your lives and your progressions of lives are improved in all possible ways. The improvements can be *medical:* a genetic defect is made normal; an injury or infection fails to occur. The improvements can be *physiological:* the bodies in your next life become stronger and healthier; the powers of all your organs are amplified. The improvements can be *personal:* you don't suffer from old misfortunes or make old mistakes. The improvements can be *moral:* your vices are weakened, your virtues are strengthened. The improvements can be *social:* you will be born into a more just society. The improvements can be *environmental:* each revised earth is slightly kinder to human animals. You will live in an environment that poses fewer threats and more opportunities. Overall, your future lives will be longer and richer versions of your life.

99. Biological transcendence

The epic of biology is the full account of the revision relation. It states that every possible life is revised in every possible way, over and over again, through all possible levels of excellence. Revision is not restricted to human lives. The entire tree of life on earth, with all its myriad instances of all its myriad species, will be revised. Every bacterium, plant, insect, reptile, mammal, primate, and human will be revised. And revision is not restricted to earthly lives: if there are alien lives on other planets in our universe, then they too will be revised, on their own improved planets. Finally, revision is not restricted to our universe: all possible lives in all possible universes will be revised.

Every life in every ecosystem in every universe will be revised. Every life sits at the root of its tree of lives; every ecosystem sits at the root of its tree of ecosystems; every universe sits at the root of its tree of universes. The recursive self-improvement of universes is a vast biological enterprise. Every life is divinely projected into its successors and into its transfinite limits. But every god is surpassed by infinitely many greater gods, every universe by infinitely many better universes, every ecosystem by infinitely many better ecosystems, and every life by infinitely many better lives. Revision is topless; it has no end; it has no goal besides its own perpetual self-transcendence.

Many older philosophies posit final goals or objectives for any human life. Plato says the ultimate goal of human life is the vision of the Form of the Good (*Republic*, 507a–511e). Plotinus says the ultimate goal of human life is the dissolution of the self in the One. Buddhism says the ultimate goal of human life is *nirvana* – the state of enlightened bliss beyond the wheel of birth and death. Christianity says the ultimate goal of human life is *the beatific vision* – the direct intuition of POG. Hegel says the progressive development of history will end in the final perfection of Absolute Knowing. Teilhard de Chardin says that evolution will end with the Omega Point.

Against those philosophies, digitalism implies that there is no final goal of life at all. For digitalists, the divine hierarchy is topless; the cosmological hierarchy is topless; the biological hierarchy is topless; and your tree of lives is topless. There is no sun above and beyond nature – nature itself is the sun. All the ascents in your tree of life (all the linear sequences of lives in that tree) are absolutely infinitely long. They are as long as the Long Line of ordinals, which has no end. Consequently, your ascents are not flights to any solar destination; they do not rise towards any finality. Your ascents will not end with any vision of the Form of the Good; nor will they end with any final union with the Plotinian One or with POG. They will not end with any final *nirvana*. You do not seek to escape from the wheel of birth and death – on the contrary, you seek to keep the wheel turning. Your deepest and most natural purpose is perpetual self-transcendence. Your perpetual self-transcendence is an ever higher flight into ever brighter *light*.

100. Revision is not resurrection

You are changing into your future ascentmates. Hopefully, your earthly life is still going strong: you are changing into some future earthly bodies. If you are, then some of your future ascentmates are in your present earthly life. However, some day you will die. After you die, the epic of biology says that you have future ascentmates in better versions of your earthly life. You have future ascentmates in your revised lives. Of course, your future ascentmates are 3D bodies in 4D lives. Any future ascentmates in your revised lives are your *better bodies*. By definition, you *will be revised* if and only if you are an earthly body and you will be a better body. The epic of biology entails that you will be revised, over and over again, in all possible ways, through all possible levels of biological excellence. You *will be* identical with every one of your future better bodies.

You might try to argue that the evolution of the concept of resurrection justifies the thesis that revision is a type of resurrection.[6] After all, Hick affirms that every resurrection life begins with "something analogous to birth" (Hick, 1976: 465). However, when some old life is resurrected into some new life, the new life continues into the old life. The biological or psychological content of the *last* stage of the old life is somehow carried over into the *first* stage of the new life. However, this does not happen in revision.

For revision, the new life is not derived from the last stage of the old life. Rather, the entirety of the new life is derived from the entirety of the old life. Revision entails that the *first* stage of your new life carries the content of the *first* stage of your old life. Your new zygote is a revised version of your old zygote. For revision, there is an approximate correlation of your old bodies with your new bodies: roughly speaking, the *n*-th stage of your next life is a revised version of the *n*-th stage of your previous life. While resurrection promises the continuity of some series of 3D bodies, revision promises the continuity of some series of 4D lives. Revision is not resurrection.

101. Revision is not reincarnation

You might try to argue that revision is some type of *reincarnation*.[7] After all, the old life and the new life are both lives of the same biological essence. So, the essence of the old life is reincarnated in the new life. It is reinstantiated in the new life. If reincarnation were to mean *nothing more than* the reinstantiation of an essence, then revision would be a type of reincarnation. Unfortunately, reincarnation means much more than mere reinstantiation. It is usually thought to entail several additional big doctrines.

The first big doctrine states that reincarnation entails some type of mind–body dualism. Essences are immaterial minds. Hence the reinstantiation of an essence becomes thought of as the *movement* of one immaterial mind from an old body to a new body. To use familiar language, reincarnation involves the *transmigration* of minds (and it makes no difference if one talks instead about the transmigration of souls or essences). The mind–body dualism of traditional reincarnation further entails that you might somehow be able to recall your past lives. It also entails that the nature of the body is irrelevant to reincarnation. Your mind can be reincarnated in any type of body. You might be reincarnated as a human of some other gender, or even as an animal or plant. The second big doctrine is that reincarnation seems to entail some type of compensatory or retributive justice (in the East, it is called *karma*; among the Neoplatonists, it is called *adrasteia*). The third big doctrine is that reincarnation is usually thought to happen here on earth. Thus a human who dies on earth might "come back" as another human or as an animal.

Revision contradicts every big doctrine entailed by reincarnation. First, revision rejects mind–body dualism and rejects every notion of transmigration. Essences are not immaterial minds; they are abstract biological programs. Since they are abstract objects, essences do not exist in space or time; hence they cannot *move* from body to body. Just as numbers do not move, so essences do not move. Since revision rejects mind–body dualism, it rejects the notion that anyone can recall their past lives. Since there is neither biological nor psychological continuity in revision, it is not possible for any brain in any revised body in any revised life to recall its past lives. Past life recall is an absurd superstition. Second, revision rejects karma. Since revision is based on the improvements of lives, karma is irrelevant. An evil life will be revised into a better life. Perhaps there are types of punishment (involving types of suffering) which can lead to improvement. If so, then revision allows evil lives to be punished in those ways. However, revision does not permit retribution or revenge. Finally, revision does not happen on our earth – it does not even happen in our universe. You will be revised in some other universe.

Consequently, revision is not reincarnation in either the traditional Hindu or Neoplatonic senses. And it is not reincarnation in the popular modern sense either. Digitalists can refer to revision as a type of reincarnation if, and only if, they can replace those old senses of reincarnation with their own new and more correct sense.

102. Revision is rebirth

Among all the classical theories of life after death, the one that seems closest to revision is the ancient Buddhist theory of *rebirth*. It was developed in Theravedic Buddhism. It is linked to the old Buddhist doctrines of impermanence and no-self. The doctrine of impermanence resembles the digitalist theory of exdurance (see Section 12). Digitalists and Buddhists both deny that there are any permanent substances that endure either through one life or across different lives. Digitalists and Buddhists both agree that there are continuities from old lives to new lives. The affinities between the digitalist concept of revision and Buddhist rebirth are nicely illustrated in this passage:

> As there is no permanent, unchanging substance, nothing passes from one moment to the next. So quite obviously, nothing permanent or changing can pass or transmigrate from one life to

the next. It is a series that continues unbroken, but changes every moment. The series is, really speaking, nothing but movement. It is like a flame that burns through the night: it is not the same flame nor is it another. A child grows up to be a man of sixty. Certainly the man of sixty is not the same as the child of sixty years ago, nor is he another person. Similarly, a person who dies here and is reborn elsewhere is neither the same person nor another. It is the continuity of the same series.

(Rahula, 1974: 34)

Of course, there are differences between the digitalist concept of revision and the Buddhist concept of rebirth. For instance, any continuities are 4D to 4D rather than 3D to 3D. However, those differences are not big enough to prevent digitalists from saying that revision is a type of rebirth. Perhaps most importantly, *rebirth* is a culturally new term with few unwanted connotations. Digitalism therefore says that if some earlier life is revised into some later life, then it is *reborn into* that later life. The epic of biology is a theory of rebirth. It is the *revision theory of rebirth*. You will be reborn, over and over again, in all possible ways, through all possible levels of biological excellence.

You *will be* your better bodies; however, you *are not* your better bodies. For identity, the tense makes all the difference: you *are not* identical with any better body; but you *will be* identical with each of your better bodies. They are your *revision counterparts*. Temporal counterpart theory supplies tensed statements with their truth-conditions. Here are some statements that are false when said by you now: "I *am* identical with some better body"; "I *am* the same person as some better body." However, if the epic of biology is true, then here are some statements that are true when said by you now: "I *will be* identical with some better body"; "I *will be* the same person as some better body"; and "I *will be* reborn." More precisely, the statement "I *will be* born again in another universe" is true when said by you now, since you have better bodies in other universes.

Since there are many ways your earthly life can be improved, you will be revised in many different ways. You will be identical with many distinct better bodies. Yet they are not identical with each other. Of course, it would be absurd to say that you *are* identical with x and you *are* identical with y but that x *is not* identical with y. But tense matters: it is entirely consistent to say that you *will be* identical with x, and you *will be* identical with y, but that x *is not* identical with y. For example, suppose John is a plumber; but John also has talents in politics and philosophy.

The epic of biology entails that he will have two distinct revision lives that actualize those talents. Temporal counterpart theory allows John to truly say "I will be a politician and I will be a philosopher." John has one better body who is a politician and another who is a philosopher. He *will be* identical with each of them, even though they *are not* identical with each other.[8] This is not paradoxical. On the contrary, it is a straightforward consequence of branching temporality.

103. The advantages of revision

At least through the mouths of the characters in his books, Nietzsche affirms that he wants to live his life over again. And if your life is worth living once, then it is worth living again (Blumenfeld, 2009). But Benjamin Franklin wants more than Nietzsche. Franklin desires *at least* the same life over again – he would like to live his earthly life over again but he would like even more to live a better version of that life. Franklin wisely prefers revision. At the start of his autobiography (1771: 1), he writes:

> were it offered to my choice, I should have no objection to a repetition of the same life from its beginning, only asking the advantage authors have in a second edition to correct some faults of the first. So would I if I might, besides correcting the faults, change some sinister accidents and events of it for others more favorable.

Franklin's desire for revision generalizes. Many people, unfortunately, have damaged genotypes. And they *shouldn't* wish for the biological or psychological continuity of lives with those genotypes. After all, it is unethical to desire to be less than the best. They *ought* to wish that their bodies were improved in every cell at the molecular level. And many people, unfortunately, have damaged earthly biographies. They ought to wish that their lives were different at every moment from conception onwards. They ought to want to have better versions of their damaged earthly lives. Fortunately, the epic of biology entails that they *will* have those better versions. It entails that more of what ought to be true for each person will be true for that person. Consider the following examples:

From conception, some people suffer from horrible genetic illnesses. Cystic fibrosis has already been discussed. But consider *Williams Syndrome* – it deeply damages the entire nervous system, resulting in profound mental retardation. Little Timmy is too mentally damaged to form any coherent desires. Using the idea that genes have some intentionality, we might say that his genome wishes for him – abstractly, his *soul* wishes for him. It wishes that his body were different, in every cell, at the molecular level, starting with his zygote. Of course, continuity can't satisfy that wish – but revision can.

From an early age, many people suffer from deep chronic illnesses. Little Cindy has *juvenile diabetes*. Her illness will negatively structure her entire life. Little Cindy says: "I wish I could have a normal childhood!" Continuity can't satisfy that wish – she will have aged. But revision can. Her better versions will be healthy – she will have a normal childhood, and a better life. So she can truly say: "When I am reborn, I will have a normal childhood. I will know what it's like to be a healthy little girl."

From an early age, many people suffer from abuse at the hands of others. Little Bobby is a victim of *fetal alcohol syndrome*. His mother was an alcoholic. He wishes he were able to have a normal life – he wishes his entire body was deeply different, and he wishes his whole life, from conception through death, was different. He wishes he were able to ride a bike, to play joyously with others, to function well in a relationship with a woman. And this entails that he wishes his mother could live much of her life over again too. These wishes cannot be satisfied by continuity. They can only be satisfied by revision. Fortunately, in the next revision universe, his mother will not be an alcoholic, and he will not be damaged. He will be healthy, and will live a better life. His desires to flourish, both physically and mentally, will be satisfied by his future revisions.

After falling deeply in love and getting married, Charles and Ann try to have children. But they are *reproductive failures*. They desperately wish they were able to have children of their own. They want fertile versions of their *earthly* lives together. They want to live their lives over again with children. Yet those desires cannot be satisfied by continuity through death. But revision actualizes the positive potentials of couples. Both of their lives will be improved together – their better versions will fall deeply in love, get married, and have children. Their desire to flourish as a couple will be satisfied.

104. Universal salvation

Salvation does not reside in any eternal substances – there are no such things. On the contrary, salvation resides in *eternal structures*. Digitalism is a *structural soteriology*. The *salvation* of any thing is an eternal structure

in which every better version of that thing exists. It is an eternal structure in which every possible positive potential of that thing is realized.
The salvation of any thing is the tree of revisions of which it is the
root. The salvation of any life is its tree of lives. Any thing is *saved* if
and only if its salvation exists. Since your tree of life exists, you are
saved. Likewise, every life in our entire earthly ecosystem is saved; our
earthly ecosystem is saved; every ecosystem in our universe is saved; our
universe is saved; every possible universe is saved; concreteness is saved.

Since the epic of theology entails that every god sits at the root of a
divine tree of lives, that epic entails that every god is saved. The creative
acts of the digital gods ensure that all things which supervene on those
gods are saved. All possible physical things are saved. Perhaps these digital gods are saviors or redeemers – yet they are not messiahs. They save
as the result of their natural actions. Salvation is the result of *nature
naturing*.

Your life will be saved by the natural action of our local god. You
do not owe any debts to our local god or to any other gods. They will
save you no matter what you do. And they save all things which they
call into being. The digital gods are compassionate, and this compas-
sion is perfect love. Digitalism entails the salvation of every possible
thing. It entails *universal salvation*. This universal salvation is the consequence of the ultimate laws of nature. These ultimate laws of nature,
which are rooted deep in its abstract background, actualize all the positive potentials of all possible things. These ultimate laws of nature are
axiological laws, ensuring that everything that ought to be true will be
true. Section 127 will give an Axiological Argument for these ultimate
natural laws.

105. Determinism and freedom

Our universe is the revision of some previous universe; our ecosystem
is the revision of some previous ecosystem; and your life is the revision
of some previous life. Just as you will have better lives in future better
ecosystems in future better universes, so you have had worse lives in
worse past ecosystems in worse past universes. Of course, better and
worse refer to greater and lesser degrees of intrinsic value (Section 74).

Your current life is a revised version of your previous life, which was
lived out in some previous ecosystem, in some previous universe, running on some previous god. Once your previous life was over, that
previous god figured out the ways to improve it. One of these ways is
the script for your current life. Your current life follows this script, which

was defined in its entirety, from your conception to your death, before our universe began. This script includes your present body-program (your present soul) as well as its inputs. It defines a life that is better than your last life. Perhaps your present life is longer than your last life; or more healthy; or you do not suffer from some injury or misfortune; or perhaps you are more virtuous in this life than you were in your last life.

Your present life is determined.[9] Your future actions are defined by a script that was written out long ago. And your present life-script was written out by a god (Chapter VI). On this point, digitalism resembles many earlier types of *theological determinism,* which said that your life was defined by God before the creation of this universe. Philosophers like Leibniz and Spinoza held this view. Theological determinism is consistent with the *compatibilist* way of thinking about freedom. Compatibilists say that your behaviors are *free* exactly insofar as they originate within your own nature. Your behaviors are *not free* exactly insofar as they do not originate in your nature. Digitalists are compatibilists with respect to free will and personal responsibility.

Your nature is your body-program; it is your soul. Your soul contains sets of inputs, states, and outputs; it contains a transition function that maps (input, state) pairs onto states; it contains a production function that maps (input, state) pairs onto outputs (see Section 28). The states of your soul can be divided into functional and dysfunctional states. Your *functional states* are those in which your body is working normally. If the production function in your soul transforms a pair of the form (input, functional state) into some output, then the production of that output was free. Your *dysfunctional states* are analogous to the error states of a computer – they are the states in which the computer is no longer working within its functional specifications. If you are tied up or otherwise caused to be paralyzed, or if you are given some drug that disrupts your normal brain functionality, then you enter a dysfunctional state. If you go without water, food, or sleep for too long, then you enter dysfunctional states. You remain in a dysfunctional state until the erroneous condition is corrected. If your soul transforms an (input, dysfunctional state) pair into some output, then the production of that output was not free.

Determinism does not preclude moral deliberation. As Leslie correctly tells us, an entirely deterministic chess-playing program can consider alternatives, rationally evaluate them, and come to a decision about which is best (2001: 139–45). For moral deliberation, the best is defined by moral criteria. Your mind, going through its states, or in a transition from one state to the next, goes through moral deliberations.

Determinism does not imply that you lack choices. At almost every moment, you surely have many options, many possible future courses of action. From moment to moment, through your deliberation, you narrow these options down to your one choice, you make your decision to do this rather than that. And to say that you *can* do otherwise means exactly that you have counterparts who do otherwise. Determinism does not preclude responsibility. Since your free actions flow from your nature, you are entirely responsible for those actions. While our local god is ontologically responsible for your life-script, no god is morally responsible for any of your actions. Determinism does not stop you from having virtues and vices. Your virtues and vices are the moral tendencies in your own nature.

When you become aware that your life is theologically determined, you may falsely come to believe that you are missing something important – that you are missing your *libertarian free will*. But since there is no such thing, you are not missing it. Or you may come to suffer from the illusion that you are not the master of your own destiny. And that false belief may trigger false emotions (such as anger, or despair). Or it may lead to laziness: you decide to let the gods do all the work. Of course, these illusory beliefs and emotional reactions are already built into your life-script. When you become aware that your life is theologically determined, you *ought* to come to see that you are the master of your own destiny. You ought to feel exhilaration that your life has been divinely planned and that you are realizing the positive potentials in your soul. Your will participates in the divine will; your life works out the divine plan for all things. But this plan entails both your salvation and that of every other possible thing. Thus besides working out your own salvation, you are working out the salvation of all other things.

106. Moral struggle

There is plenty of room for moral effort in digitalism. You are an autonomous agent, and the amount of effort that you put into your life matters for your success. You are not passive. It is easy to formalize your moral struggle by adding your amount of moral effort to the rules in your soul. On this extended formalization, your moral effort enters the antecedent of each if-then rule in your soul, as an input from your body into itself. Thus (1) if you're faced with temptation and you put up no resistance to it, then you succumb and you indulge; but (2) if you're faced with temptation and you put up lots of resistance, then you triumph over temptation and you don't indulge.

Your self-updating (which follows the rules in your soul) is more or less intense. Digitalism entails that if you fight your own tendencies to evil with greater intensity, then you have a greater chance of winning over them; however, if you surrender to those tendencies, then you lose. According to compatibilist accounts of personal responsibility, you alone are responsible for yourself. The fact that your life is a part of the life of some god does not decrease your responsibility for yourself. Each cell in each body is responsible for itself; each part of each god is responsible for itself. You cannot blame any god for your behavior any more than you can blame your society, your genes, your upbringing, or your brain. As long as you are not coerced, you do what you do because of your own nature. And it matters not that your nature was designed by some previous divine nature. Your self-updating originates entirely within yourself. It matters not that it supervenes on some present divine activity. Your life is a part of the process of your local god. All this means is that you are not ontologically foundational – you lack aseity. It does not mean that you are unable to exert any moral effort. You ought to construct the best life you can; by doing so, you advance your own progression of lives.

Of course, if you surrender to evil, then you will still make progress. But your progress will be slower and it will involve more suffering. Suppose you refuse to make any moral effort to resist the temptations of vicious drugs. You become an addict; you end up diseased and in jail; your earthly life is one of intense suffering. This suffering is a part of the suffering of your local god – and that god will redeem it in every improved version of our universe, which contains some improved version of your life. But improvement is a slow and gradual process. You will suffer less in every one of your next lives – and yet you will still suffer *almost as much*. Your addiction will be slightly less severe; your time in jail slightly less terrible; your disease slightly less painful. Digitalists affirm that all the darkness of human life will *eventually* be turned into light. But only slowly. It is up to you to optimize your ascent. If you fail to do this, digitalism provides you with hope. But it does not provide you with any excuses. You are responsible for yourself.

VIII
Superhuman Bodies

107. Optimized universes

As ever greater gods make ever better universes, *optimized universes* appear. Each optimized universe is a self-organizing process. Self-organization in optimized universes is more intense than in lower-level universes like ours. It is more common, more durable, more diverse in its forms, and it produces greater heights of complexity. To describe optimized universes, it is necessary to extrapolate. It might be objected that such extrapolation is too speculative – it is too much like science fiction. Here the correct reply is that philosophy ought to take science fiction more seriously.

Among all the optimized universes, it will be helpful to focus on a specific example, namely, *Universe Epsilon*. This is an optimized version of our universe. Its evolution begins with something like our Big Bang. Optimized physics resembles our physics: particles form atoms; atoms form molecules; clouds of gas form stars; solar systems form around stars. Solar systems provide resources for life – stars provide energy and their satellites provide other resources. One of the most important resources needed by life is *surface area*, and optimized solar systems provide lots of it. On the surfaces orbiting optimized stars, chemical and biological evolution take place.

The optimized solar systems in Epsilon are not usually orbited by planets. Consider the Epsilon-counterpart of our solar system. It is surrounded by concentric shells of small planetoids, which are tightly coupled by local gravity. The planetoids in any shell vary in size, shape, and material composition. They have mountains and valleys, deserts and oceans, and tropical and frigid zones. Each shell is bathed in its own atmosphere. Within the atmosphere, the planetoids are linked by bridges of gas and liquid. The total surface area of a shell of planetoids is enormously greater than that of any planet.

Gravitational forces and solar winds generate enormous flows of energy and matter in the shells. These flows drive optimized chemistry, which is similar to our chemistry. Complex chemical reactions cover the surfaces of the planetoids. And they spread across the bridges. Every shell of planetoids is an enormous chemical reaction in which biological evolution takes place. Optimized biology resembles earthly biology. Living things are still chemically realized – but the chemistry is better for life. The chemical reactions in optimized living systems are more reliable and efficient. They are better able to realize more diverse and more highly functional biological machines.

On some of the planetoids, life evolves. The evolution of optimized life is similar in some ways to the evolution of life on earth. It rises through a similar sequence of types. But it is much richer. Since the planetoids are linked by bridges of material, it is possible for bird-like organisms to fly, or fish-like organisms to swim, from one planetoid to another in the same shell. Every shell becomes an enormous ecosystem filled with trillions upon trillions of flourishing species. All earthly species have optimized versions in these ecosystems. And optimized ecosystems will be filled with hybrids and chimeras and beasts that are merely mythical here. There are optimized butterflies, horses, unicorns, dragons, and so on. But here we are mainly concerned with optimized humans. Of course, some optimized humans may evolve into nonhuman organisms.

Eventually, hominid creatures very similar to earthly humans appear in these optimized ecosystems. They evolve into *optimized humans.*[1] And, just as earthly humans sail from island to island, so optimized humans sail from planetoid to planetoid. All the humans that were in our universe have optimized counterparts in Epsilon. You and I have optimized counterparts there, as do Deucalion and Pyrrha. According to digitalism, right now you can truly say: "I will be an optimized body, I will live an optimized life."

108. Enhancing your own genes

An optimized human, like every kind of human, begins with the fusion of optimized sperm and egg. It starts as an optimized zygote, which encodes its genotype. Any human genotype contains many *genetic roles*, which are occupied by DNA strands coding for proteins or functional RNA. Each role has many possible occupants (all within the human species). The possible occupants of any genetic role are its *alleles*.

The alleles in any genetic role lie on a spectrum from subnormal, through normal, to supernormal. Subnormal alleles lead to genetic diseases. Cystic fibrosis, for instance, is caused by a subnormal allele of the CFTR gene. Supernormal alleles lead to organ performance above the statistical average. For example, the ACE gene has two alleles (I and D); the I allele is associated with superior athletic endurance and superior performance in general at high altitudes (Woods et al., 2002). Of course, most abilities are associated with complex networks involving many alleles of many genes.

An optimized genotype does not suffer from any genetic defects. All optimized alleles are at least normal. There are no genetic diseases in optimized bodies. An optimized genotype contains much more information than an earthly genotype. For any human organ, an optimized genotype contains the blueprint for every normal or supernormal version of that organ. For all organ-blueprints, an optimized genotype contains all the alleles varying from normal to supernormal. It can switch them on and off.

Any earthly genotype is dynamic and adaptive. For example, the VDJ genes in immune B-cells and T-cells adapt to new challenges (Section 25). The immune system learns at the genetic level by running evolutionary algorithms on those cells – it breeds new genes. Of course, this adaptive power preserves the human genotype. As your immune system learns, you remain human. All changes stay within the range of human genetics. The regulatory logic of the immune system can be artificially extended to other gene complexes. Earthly humans may soon use artificial techniques (such as gene doping) to temporarily change genes for enhanced athletic or cognitive performance. For example, before an endurance contest, an athlete might artificially modify their ACE genes to the I allele-form (Alvarez et al., 2000). But optimized humans have the *natural* ability to change their patterns of gene expression to meet their current challenges. They can train their genotypes like we train our muscles. Their genotypes are fully adaptive.

An optimized genotype contains extra regulatory circuits. These extra circuits control other gene complexes in the same way that the regulatory circuits of the immune system control the VDJ genes in B-cells and T-cells (Benjamini et al., 1996). Just as the immune system can be made stronger by exposure to a challenge, so too any organ in an optimized human can be made stronger by exposure to challenges. For example, an optimized genotype contains a regulatory circuit that controls the genes for athletic performance. As an optimized human goes through increasingly difficult endurance challenges, this artificial circuit changes its ACE genes to the I allele-form. As the challenges fade, the circuit changes the ACE genes back to their original alleles (which, of course, might have been the I forms in the first place). An optimized genotype responds to repetition (to training and practice). As demands are made on an organ, the genotype changes it to meet them. An optimized genotype can functionally extend human organs. But it does not add any nonhuman organs. It can change a human eye into one that can see into the ultraviolet; but it cannot add feathers, wings, gills, fins, scales, or tentacles.

An optimized human starts with an optimized zygote, which carries an optimized genotype. An optimized genotype realizes an optimized body-program – it realizes an *optimized soul*. All the optimizations of any soul are in its *essence*.[2] The essence of any human soul is closed under optimization. And optimized souls guide the growth of optimized human bodies. After conception, an optimized human grows through cell division to maturity much like an earthly human. Its optimized cells are improved versions of earthly cells. As the optimized body grows, its cells and organs fill out the standard human body-plan (the human *bauplan*). The human body-plan is a definite arrangement of definite types of organs. All possible humans share this essential body-plan.[3] For example, humans are essentially bipedal (Stanford, 2003). The human body-plan includes certain organs and not others: a human has arms but not wings; it has lungs but not gills; eyes but not antennae. Although the human body-plan specifies a fixed framework of organs with specific functions, it does not constrain their functional parameters. They can grow ever more powerful. A more powerful eye is still an eye. Your better bodies will have organs whose functions are extrapolations of the functions of earthly humans.

109. Bodies with the best human powers

An optimized human can do whatever *any* earthly human can do. Any optimized organ of some type is as good as the best earthly human organ of that type. For any task F, if there is any earthly human organ that can do F, then any corresponding optimized human organ can do F. An optimized human organ of some type is *universal* relative to the whole class of human organs of that type. But optimized humans are functionally superior to *all* earthly humans. All earthly humans suffer from *species-level design defects*. These are parts of our common evolutionary heritage. The present information about these design defects is taken mainly from Olshansky et al. (2001) and from the fascinating book *Quirks of Human Anatomy* (Held, 2009). Optimized humans do not suffer from any of these species-level design defects – they are fully corrected. Optimized humans necessarily have the same types of physiological systems as earthly humans. These are as follows:

The Skeletal System. An optimized human has an optimized skeletal system. An optimized skeleton is as strong as the best earthly skeleton – it does not suffer from any earthly human defects. But even the best earthly skeletons are suboptimal. The earthly skeleton suffers from many species-level design defects. The skull has sinuses whose poor drainage makes them highly susceptible to serious infections. Although sinuses make the skull lighter, the same effect can be achieved by using lighter bone. Optimized humans lack sinuses. Our jaws are too short to hold all our teeth. The most problematic are the wisdom teeth. While some earthly humans lack wisdom teeth, all optimized humans lack them. The earthly spine and knee are poorly engineered. Optimized humans have better spines and knees. Some earthly humans (about 1 percent) grow cervical ribs (neck ribs). They are useless and sometimes cause medical troubles. Optimized humans lack the genes that grow these ribs. Earthly embryos grow tails (between four and seven weeks). The tail is useless, and a special set of genes exists to remove it. All that remains is the coccyx, made of several fused vertebrae. Although it is a single bone, there is a muscle (with an associated nerve and blood supply) to flex it. Earthly humans have tail-potentials which they *contradict* by crippling their actualizations. Optimized humans either entirely lack those potentials or else they fully actualize them. If actualized, then they have tails which are fully and appropriately human. Optimized skeletons are as good as human skeletons can be.

The Epidermal System. An optimized human has an optimized epidermal system. Their skin is as good as the best human skin. But even the best earthly human skin is naturally suboptimal. It burns when exposed to too much sun and it is easily penetrated by the disease-carrying bites of pests. Optimized human skin does not suffer from those defects. Earthly human skin contains a vestigial hair system. We have very little hair and our hairs are very small. Yet each hair is controlled by a muscle (the *erector pili*) and an associated nerve. For other animals, the hair system is used to regulate temperature and to express emotions (like fear). For earthly humans, this system produces only goose bumps (*cutis anserina*). An optimized human has a useful system of hair controllers. Your optimized bodies will use their hair to express a range of emotions.

The Sensory System. An optimized human has optimized sensory systems, which suffer from no earthly defects. Their sense organs are as good as the best human sense organs. For any sensory task S, if there is any earthly human sense organ that can perform S, then any optimized sense organ can perform S. Sadly, even the best human sense organs are suboptimal. They suffer from various species-level (design) defects.

Consider your eyes. The structure of the eye seems to be suboptimal. The retina is inverted. The photocells are installed backwards (they point away from the light towards the brain). The nerves that come from the photocells run over the top of the retina and thus block the light. Since the nerves have to go through the retina, there is a blind spot. Since the photo cells cannot be securely attached to the rest of the eye, the retina is easily detached. An optimized human has a retina with photocells pointing towards the light (a design found in squids and octopi). Although only three muscles are needed to control the eye, the earthly human eye has six. An optimized human has a better optical control system, either using the extra muscles for better focusing or dispensing with them.

Consider your ears. The earthly ear has three muscles attached to it, which have their own nerves and blood supply. For other animals, these muscles help the ear rotate towards sound. Since human ears are attached flat to the head, these muscles are useless. An optimized human has ears that can rotate and thus uses these muscles. Some blind earthly humans navigate by echolocation (Stoffregen & Pittenger, 1995). Optimized ears are better suited to echolocation. And optimized vocal systems can produce louder tongue-clicks for echolocation. Or perhaps optimized voice boxes will be able to project regular sonic pulses. Earthly humans have small vomeronasal organs. These are lined with chemoreceptors that were probably once used to detect pheromones. An optimized human has functional vomeronasal organs and emits pheromones to communicate.

The Computational System. An optimized human has an optimized nervous system, which has no earthly defects. Its nervous system is as good as the best human nervous system. For any neural task S, if there is any earthly human nervous system that can perform S, then any optimized nervous system can perform S. The brain of an optimized human can do whatever any human brain can do. And some human brains have extreme powers. For example, there are cases of nearly exact autobiographical memory (Parker et al., 2006). The brains of human savants typically have extraordinary powers in limited domains (Treffert & Wallace, 2004; Pring, 2005). These powers are almost always somehow based on supernormal memories for concrete detail. For example, the savant Kim Peek "began memorizing books at the age of 18 months, as they were read to him. He has learned 9,000 books by heart so far" (Treffert & Christensen, 2005: 108). The powers of savants are typically based in the right hemisphere. They include exceptional talents in music, art, various kinds of calculation, mechanical aptitude, and spatial reasoning. They are often accompanied by severe deficits in skills based in the left hemisphere. Savants thus typically lack linguistic and conceptual skills (such as abstraction skills). A savant may memorize a book but know neither what it means nor how to make abstract associations from it. An optimized brain has savant-level skills without the deficits.

Sadly, even the best human nervous systems are suboptimal in many ways. They suffer from various species-level (design) defects. Even the best earthly brains suffer from many design errors. Although the eyes are close to the front of the brain, the visual processing areas are in the back of the brain. They should be closer to the eyes for faster processing. The neural wiring of the larynx is suboptimal. The recurrent laryngeal nerve originates from the spinal cord in the neck (it is a branch of the vagus nerve). It runs down into the chest, passes under the aorta near the heart, then comes up to the larynx. A better design is to have the nerve run right from the spinal cord to the larynx. An optimized human has a more efficiently wired and faster nervous system.

All optimized humans have endocrine and immune systems that are as good as the best human endocrine and immune systems. It is likely that the immune system can be improved in many ways. An optimized human has a more carefully engineered and functional immune system. It does not suffer from autoimmune diseases. It does not make the mistakes that lead to allergies and asthmas.

The Motor System. An optimized human has an optimized motor system, which lacks all earthly defects. Its muscles are as good as the best human muscles. They are as strong, fast, and efficient as the best human muscles. For any motor task S, if there is any earthly human motor system that can perform S, then any optimized motor system can perform S. Earthly athletes use various procedures to improve motor performance. These include training, drugs, and genetic alterations. For example, an athlete can get erythropoein (EPO) injections; or (someday soon) the athlete can get an extra EPO gene. An athlete can take steroids or other drugs to enhance muscle size or function. An athlete can be genetically altered to make larger muscles (by adding the gene for mechano growth factor). An optimized human can naturally do whatever these artificially improved humans can do. As expected, the earthly human motor system suffers from species-level design defects. Many earthly humans have useless muscles (like the subclavius, palmaris, and pyramidalis muscles). They are missing in some earthly humans and in all optimized humans. All other muscular design defects are optimized.

The Metabolic Systems. An optimized human has an optimized metabolic system, which does not suffer from any earthly defects. It is as good as the best human metabolic system. Its digestive, respiratory, and circulatory systems are as good as the best human systems of those types. For any metabolic task S, if there is any earthly human metabolic system that can perform S, then any optimized metabolic system can perform S. But even the best earthly human metabolic system suffers from species-level design defects. For example, vitamin C is essential for human metabolism. While earthly humans have a gene for making vitamin C, that gene is broken (hence we suffer – and suffer horribly – from scurvy if we don't get enough vitamin C from our diets). An optimized human has a working vitamin C gene. As another example, consider the pharynx. The pharynx is a passage used in the earthly body for both ingestion and respiration. This design leads to the risk of choking. One might argue that the pharynx provides a valuable redundancy (we can breathe through our mouths if our nasal passages are clogged). If a better design is possible (as seems likely), then an optimized human has that better design. The appendix appears to be a vestigial organ (although it may play some role in the immune system). Still, it puts the body at risk for appendicitis. An optimized human either has no appendix or one that performs the immune function without the high risk of appendicitis. An earthly human male has a urethra that passes through the prostate. If the prostate swells (which it almost always does), the urethra becomes constricted, causing troubles. An optimized urethra goes around the prostate. It is likely that there are many other optimizations for human digestive, respiratory, and cardiovascular systems.

The Reproductive System. An optimized human has an optimized reproductive system, which lacks all earthly defects. It is as fertile and reliable as the best human reproductive systems. The earthly reproductive system suffers from species-level design defects. The male reproductive system is a variation on the female system. Males have nipples and the capacity to produce milk (if given hormones). Males also have a degenerate uterus hanging off the prostate gland (the prostatic utricle). One might suggest that optimized humans are hermaphroditic. But male and female bodies differ in more than just their reproductive systems (for instance, their nervous, endocrine, immune, and motor systems are different). Optimization does not produce a fully hermaphroditic human. But it can go part way. It would be a great biological advantage if a man were able to nurse an infant under emergency conditions. Since this capacity seems available but frustrated in earthly men, optimized men gain this ability. No doubt this sounds weird. But it is hard to imagine a father who would prefer to watch his child die from malnutrition than to be able to nurse his child back to health. Finally, menstruation is costly for women – they lose considerable amounts of iron. And pregnancy and birth are costly and risky. Any design defects in the female system will be revised to excellence in optimized women.

The Repair System. After maturity, optimized bodies do not suffer from any subnormalities. Many subnormalities cause premature aging. For example, high blood pressure weakens the blood vessels and leads to strokes; poorly tuned genetic regulatory networks lead to cancer. Many other subnormalities lead to deficiencies that cause many earthly humans to lead shorter than normal lives. Optimized humans live maximally long human lives – they live as long as the longest earthly humans live (say, close to 120 years). Optimized cells have optimal molecular repair powers. For any organ, and for any protective or healing power, if there is some earthly human organ that has that protective or healing power, then every optimized organ has that power. For any event that can happen to an earthly human, if there is some earthly human body that can survive it, then every optimized human body can survive an event of that type.

Although optimized humans may live as long as the longest earthly humans, their lifespans remain finite. Optimized humans age and die. And their optimized ecosystems die as well. Any optimized universe eventually becomes sterile and structureless. And the gods that run them eventually cease to operate. Fortunately, those gods are surpassed by greater gods; but those greater gods make idealized universes, which contain idealized ecosystems, idealized lives, and idealized bodies.

110. Idealized universes

As ever greater gods make ever better universes, *idealized universes* appear. An idealized universe (like any universe) is a self-organizing process; but self-organization is more intense in idealized universes than in optimized universes. Among all the idealized universes, it will be helpful to focus on *Universe Zeta*. It is an idealized version of Universe Epsilon (and, therefore, it is an improvement of our universe).

Idealized physics is an extrapolation of optimized physics. But it is a long way from earthly physics. Universe Zeta begins with an event that fills it with primeval energy. This energy condenses into particles, which serve as the building blocks for more complex structures. Zeta begins with a hot rich fluid medium in which these particles swarm and interact. At some places, they condense into denser, cooler, and more solid structures; at others, they remain thinner, hotter, and more dynamic.

There are no solar systems in Zeta – no stars, no planets. What does Zeta look like? Picture a 3D connect-the-dots network whose links and nodes are made of solid stuff, much like earthly rock. These nodes may be as small as our earthly moon or as large as entire earthly solar systems. Nodes are always connected to many other nodes by solid links. The arcs that join them may short and thin; or they may be light years long and trillions of miles in diameter. The surfaces of these solid nodes and links resemble those of planets: they are covered with mountains and valleys, deserts and oceans, jungles and ice. They are surrounded by atmospheres, and these atmospheres fill the spaces between the solid filaments. As you rise into any atmosphere, it becomes hotter and brighter. A connect-the-dots network of radiant energy occupies the spaces between the filaments of solid stuff. The nodes of this luminous network pulsate with celestial energy. The arcs are pulsating streams of hot bright plasmas. Parts of the solid network are illuminated and darkened as these arcs and nodes of energy pulsate from noon to night.

Idealized chemistry is an extrapolation of optimized chemistry. Optimized structures have idealized analogs. These analogs are still recognizable from our earthly point of view; there are idealized counterparts of our chemical elements (idealized hydrogen, carbon, oxygen, and so on). Chemical structures form living cells and organisms. The chemistry is better suited for life. The chemical reactions in idealized living systems are more reliable and efficient. They are better able to realize higher-level biological machines. Life evolves on the surfaces of the solid network; it evolves in the liquids that run over those surfaces; it develops in the atmospheres. The evolution of idealized life is still somewhat like that of life on our earth. But it is enormously richer and more intense.

An idealized universe contains an enormous diversity of life. It contains idealized versions of all earthly species that have ever existed. It contains idealized versions of all possible hybrids and chimeras and mythical creatures. Eventually, hominids evolve in this idealized universe. Some of these hominids evolve further into earthly humans; then into optimized humans; then into *idealized humans*. Of course, some earthly, optimized, or idealized humans may evolve into nonhuman organisms. But we are not concerned with them here. Here we are concerned only with humans.

All idealized humans are physical animals living on the solid surfaces in idealized universes. Revision entails that all the optimized humans who lived in Universe Epsilon have idealized counterparts in Zeta. And, by transitivity, every human who lived in our universe has an idealized counterpart in Zeta. They are idealized ascentmates. Zeta contains idealized versions of you and I as well as idealized versions of Deucalion and Pyrrha. All idealized humans have idealized lives. Digitalism entails that right now you can truly say: "I will be an idealized body, I will live an idealized life."

111. Universal genetic logic

An idealized human necessarily starts from the fusion of idealized sperm and egg. It begins as a one-celled human – as an idealized zygote. The zygote encodes an idealized genotype. An idealized genotype contains an adaptive blueprint for every organ in the human phenotype. For any organ in the human phenotype (in the human body-architecture), an idealized genotype contains a *carbon-universal blueprint*. The carbon-universal blueprint for an organ can generate an organ whose functionalities are carbon-universal. If the functionality of an organ is carbon-universal, then it can do whatever any carbon-based version of that organ can do. An idealized human can perform any *human* body-function as well as the *best earthly organism of any species*. An idealized human is as good at any task as any optimized carbon-based organism. Thus idealization is universalization over all carbon-based optimized organs.

As an illustration of carbon-universality, consider eyes. Since the human body-plan has eyes, an idealized genotype contains a carbon-universal blueprint for them. What does this mean? Consider all the types of eyes that are found in earthly species. There are insect eyes, reptile eyes, and so on. They all have different functional powers – they are sensitive to different parts of the spectrum, they have different focusing and resolving powers, and so on. The idealized genotype contains an eye-blueprint for *every type* of *optimized* eye that can be built out of protein. It contains the blueprints for optimized human eyes and optimized eagle eyes. If an idealized human needs the powers of an eagle eye, the idealized human eye morphs into an eye with those powers. The lowly mantis shrimp has eyes with extreme powers. If necessary, the idealized human genotype can convert an idealized human eye into an eye with the optical powers of mantis shrimp.

An idealized human genotype is dynamic and adaptive like an optimized human genotype. For every organ, it includes the relevant genetic networks of other organisms. It can switch them on and off as needed. When an idealized human is challenged to perform some function, it activates the relevant genetic networks to modify its organs. For example, if it needs to see into the ultraviolet, it calls upon the genetic networks needed to rewire some of its photoreceptors to become sensitive to ultraviolet light. Suppose an idealized human needs to generate energy from light – it needs to perform photosynthesis. Plant cells have molecular structures that can perform photosynthesis. The genetic networks for these structures are contained in the skin-cell-blueprints of idealized humans. So, when they need to perform photosynthesis, they call upon those genetic networks. Their skin cells construct the needed molecular machines and the photosynthesis begins. The addition of the ability to do photosynthesis does not change the human body-architecture. Skin cells, which humans already have, are modified internally to gain new functionality.

An idealized human starts with an idealized zygote, which carries an idealized genotype. An idealized genotype realizes an idealized body-program – it realizes an idealized soul. All the idealizations of any soul are in its *essence*.[4] The essence of any human is closed under idealization. And idealized souls guide the growth of idealized human bodies. After conception, an idealized human grows through cell division just like an earthly or optimized human. It grows through a similar series of cellular doublings to maturity. But its idealized cells are ideal versions of earthly cells. For any human type of cell C, and for any cellular task T, if there is some earthly species whose cells of type C can do task T, then an idealized human cell of type C can also do T. For instance, an idealized human nerve cell can do whatever any protein-based nerve cell can do. If a nonhuman nerve cell can perform some task, then an idealized human nerve cell can perform that same task. Idealized human cells are carbon-universal cells. But their highest-level genetic regulatory mechanisms remain human – it is their *human* genetic logic that decides what they do.

112. Bodies with the best earthly powers

An idealized human has idealized human organs – it has idealized eyes, hands, arms, and legs. It has an idealized heart and an idealized brain. It has all the organs in the essential human body-plan. It excludes non-human organs that violate that plan (for instance, it has no gills, fins, beaks, wings, and tentacles). And yet it may include nonhuman organs that are consistent with that body-plan. An idealized human organ of some type is universal relative to the whole class of protein-based organs of that type. For any task F, if there is any protein-based organ (of any species) that can do F, then the corresponding organ of an idealized human can do F. Idealized human eyes can do anything that any protein-based eyes can do (like polarizing light across the spectrum). Idealized human brains can do anything that any protein-based brains can do. Idealized humans necessarily have all the same physiological systems as earthly humans. They are as follows:

The Skeletal System. The skeletal systems of idealized humans converge to carbon-universality. Their joints are stronger and more reliable—they can bear larger forces more frequently without degrading. Their bones are as light as those of a bird and as strong as those of a cat. Just as cats can survive falls at their terminal velocities (Diamond, 1989), so idealized humans can survive falls at their terminal velocities. Nevertheless, every idealized human has a human skeletal structure (rather than a bird or cat skeleton). Of course, there are various human ways to extend the earthly human skeletal structure. Consider Vishnu, who is depicted with an extra pair of human arms. Vishnu has an extended version of the earthly body-plan – yet that extension remains human.

The Epidermal System. The epidermal systems of idealized humans converge to carbon-universality. For any protein-based organism, if its surface can do something, then the surface of an idealized human can do it too. Idealized skin is as supple as earthly human skin yet as tough as leather or scales. It is not easily penetrated. Idealized humans need no clothes. As an illustration of nonhuman organs which are consistent with the human body-plan, consider *chromatophores*. These are skin cells which can change their color. They are found in cuttlefish and chameleons. But adding them to idealized humans does not violate the architecture of humanity. You would *recognize* such a body as human. The patterns displayed by such idealized skin can be deliberately controlled, so that it displays information about the body (moods, desires, preferences, thoughts, social affiliations, and so on). Yet the control of this skin-display, this self-tattooing, is an artistic exercise requiring training and skill. Each idealized body aims to maximize the beauty of its skin-display. Idealized humans have hair, whose growth rate and color are deliberately controllable. An idealized human can be hairless or grow intentionally designed patterns of colored hair all over its body. Growing such patterns requires talent and training – it is an artistic exercise. Idealized skin potentially contains photoreceptors like those in plants.

An idealized human can perform photosynthesis, and convert celestial energy (or any radiant energy) into useful metabolic energy.

The Sensory System. The sensory systems of idealized humans converge to carbon-universality. If the growth of a sense organ is consistent with the human body-plan, an idealized human can grow and use that sense organ. Many animals sense the geomagnetic field of the earth (Walker et al., 2002). An idealized human can sense magnetic fields. An idealized human eye can do whatever any eye made of protein can do. It can perform any protein-based optical function. For any visual operation V, if there exists an eye made of protein that can do V, then an idealized human eye can do V. An ideal eye has ideal visual acuity. Normal visual acuity is 20/20. Laser surgery can correct aberrations in the cornea to improve contrast sensitivity and visual acuity (Liang et al., 1997; Guirao et al., 2002; Yoon & Williams, 2002). It is estimated that laser surgery can enhance visual acuity to 20/5. Laser surgery produces an ideal cornea. But an idealized human eye can do whatever any eye of any optimized earthly organism can do. It can be a human eye with the functional powers of an eagle's eye. It becomes such an eye through training.

An idealized eye has ideal color perception. Cone cells are sensitive to certain radiant frequencies (colors). Earthly humans have *trichromatic* vision. We have three distinct kinds of cone cells in our retinas. These have sensitivity peaks for the colors blue, green, and red. The earthly human visual spectrum thus runs roughly from 400 nm to 700 nm. Many animals have *tetrachromatic* vision (Honkavaara et al., 2002). They have four distinct kinds of cone cells, with sensitivity peaks for the colors ultraviolet, blue, green, and red. The tetrachromatic visual spectrum thus runs from about 320 nm to 700 nm. The cone cells in tetrachromatic retinas also use small oil droplets to filter the light. The filtering increases color saturation. It thus increases color discrimination and color constancy. We could use genetic engineering to make tetrachromatic human eyes. But an idealized human has built-in tetrachromatic vision. Tetrachromatic human vision is an idealization that is consistent with the human architecture – a tetrachromatic eye is still an eye. The eyes of the mantis shrimp have powerful achromatic polarizers across the visible spectrum (Roberts et al., 2009). An idealized human eye has equivalent polarization powers.

The Computational System. The computational systems of idealized humans converge to carbon-universality. For any cognitive task C, if there is any protein-based organism whose nervous system can do C, then the nervous system of any idealized human can do C. Idealized brains are biologically universal computers. If any species has neural circuits specialized to perform some task, the idealized brain has such circuits.

Some kinds of birds are extremely intelligent (Emery, 2006). These are the corvids (such as crows and ravens) and parrots. Corvids and parrots are in many respects as intelligent or nearly as intelligent as nonhuman primates. Apparently, the structure of corvid and parrot brains differs from that of primate brains – the architecture of the primate brain is not the only way to achieve high intelligence. Corvid and parrot brains are very small compared with primate brains. Corvids and parrots seem to pack a lot of intelligence into a small brain. The cognitive architecture of the avian brain may be more efficient than that of the primate brain. Still, we know very little (circa 2013) about either bird brains or primate brains. But if bird brains are more efficient, then idealized humans may have much of their brain architecture organized like those avian brains. And the avian alternative to the primate brain-structure suggests there may be other alternatives. An idealized human has the maximally powerful protein-based architecture for *human* cognition.

The immune systems of idealized humans converge to biological universality. A fully idealized human is such that for any germ G, if there is any earthly organism whose immune system can resist and defeat infection by G, then the immune system of the fully idealized human can resist and defeat infection by G.

The Motor System. The motor systems of idealized humans converge to carbon-universality. For any motor task M, if there is any earthly organism whose motor system can accomplish M, then the motor system of any idealized human can accomplish M. Of course, idealized motor systems are restricted to the structures found in the earthly human phenotype (such as arms and legs, hands and feet). Idealized humans don't have wings or fins to control. They don't have four legs like horses or cheetahs. Idealized humans are not hybrids or chimeras like centaurs. But idealized human legs can do whatever horse legs or cheetah legs can do. They can go fast. All muscles work faster and more efficiently. So the idealized body can run faster and farther. Arms and legs can move at the highest possible speeds for protein. Fingers (and even toes) have maximal dexterity with maximal ranges of movement. The motor system also includes the vocal system. For an idealized human, the vocal system is biologically universalized. For universal communication, it can make any sound that any carbon-based organism can make.[5]

The Metabolic System. The metabolic systems of idealized humans converge to carbon-universality. The cells of an idealized human can carry out any metabolic task that can be done by any protein-based cell. If any earthly cell that can gain energy from some chemical reactions, then idealized human cells can gain energy from analogous chemical reactions in idealized universes. But idealized humans can also gain energy from light or other radiant sources. The metabolic organs include the respiratory, digestive, and cardiovascular organs. The idealized versions of these organs are carbon-universal.

All earthly and optimized humans have mammalian *tidal* respiratory systems. They work like this: inhalation pulls air into the lungs, then exhalation pushes it out. This pull–push system entails that the air in the lungs is always partly stale. A more efficient system ensures that air constantly *flows through* the lungs (in particular, through the gas-exchanger parts of the lungs). Birds have a *flow-through* system (Powell, 2000). They use air sacs in such a way that fresh air flows through the lungs during both inhalation and exhalation. The avian flow-through system is about ten times as efficient as the tidal system. An idealized human uses something like the avian respiratory system. The incorporation of this type of nonhuman system does not violate the human body-plan. Superior lungs remain lungs. A body with such lungs would still be recognizably human.

The Reproductive System. The reproductive systems of idealized humans converge to biological universality. Each idealized human has the capacity to be either fully male or fully female. It is innately a hermaphrodite. Its gender is developed rather than hardwired. It always has the capacity to change (like Tiresias) from one sex into another. But this conversion, like every metamorphosis, requires work and takes time.

The Repair System. Idealized humans work reliably. For any organ, and for any protective, or healing, or regenerative power, if that power is within the range of any protein-based organism, then every idealized human organ has that power. For any event that can happen to an earthly organism, if there is some earthly species that can survive such an event, then every idealized human can survive such an event.

An idealized human can regrow lost limbs. Amphibians have this power, as do other earthly organisms (Brockes, 1997; Pearson, 2001). When an idealized human loses a limb, the blood supply to it quickly shuts off. The genes that grew it are reactivated, and the limb regrows. An idealized human can live as long as any protein-based organism can live. Since bristlecone pines can live for thousands of years (the longest known lived 4900 years), idealized humans can live that long. However, more generally, an idealized human can live as long as the entire earthly biological process. Every cell in any earthly human body is a stage of a biologically continuous process which stretches back to the first earthly cell. It is a stage in a living process which has been going on for well over 3 billion years. Any idealized life can last at least that long. As old cells in idealized bodies are replaced with new cells in ideal ways, those bodies persist even longer.

Although idealized humans may live as long as protein-based life itself, their lifespans remain finite. They are universalizations of protein-based biochemistry. Even within ideal biochemical systems, errors accumulate – even idealized humans age and die. And the idealized ecosystems in which they live likewise die. Any idealized universe eventually becomes devoid of life, then devoid of all structure. Idealized universes decay into noisy waste. And the gods that run them eventually cease to operate. Fortunately, those gods are surpassed by greater gods; but those greater gods make extended universes, which contain extended ecosystems, extended lives, and extended bodies.

113. Extended universes

As ever greater gods make ever better universes, sequences of *extended universes* appear. It will be helpful here to focus on one series of extended universes. This *local series* starts with the last idealized universe – it starts with Zeta. Every universe in this local series is surpassed by its next extended universe. The self-organization of every next extended universe is twice as intense as that of its previous universe.

Any extended universe begins with an initial event that fills it with particles of *computronium*. Any particle of computronium is a universal computer with some finite memory and finite speed. From each extended universe to the next, the speed, power, efficiency, and reliability of these particles doubles. And the size of these particles falls by half. Particles of computronium run different programs, which give them different physical natures. The differences in these natures drive the processes in any extended universe. As the indexes of the extended universes rise, old physical concepts become less meaningful. Matter grows ever more *subtle* – it is an increasingly universal plastic. The particles of computronium become ever more *spiritual*.

At the start, all the particles in any extended universe are seeded with initial simple programs. But these programs are all capable of recursive self-improvement. Hence they evolve into more complex programs. Particles of computronium form informational bonds – channels through which they exchange signals. And, just as any particle is capable of recursive self-improvement, so any network of particles is capable of recursive self-improvement. Networks become ever more complex. The evolutionary powers of these networks drive extended chemical evolution. Extended chemistry is a richer variant of idealized chemistry. Idealized structures have extended analogs. Chemical structures form extended cells and organisms. Hence extended life evolves.

Organisms in extended universes are realized in computational regularities closer to the bottom-level regularities of the universe. Living systems are made out of a biologically universal plastic. This universal bio-plastic can run any biological program. Extended ecosystems are ever more diverse. They include chimeras, all the mythical beasts we have ever imagined, and many we have not. Hominids like earthly humans emerge; they evolve into optimized humans; they evolve into idealized humans; and eventually they evolve into *extended humans*. Of course, many humans will evolve into nonhumans. But here we consider only humans. All earlier humans have counterparts in the local series. You and I have extended counterparts, as do Deucalion and Pyrrha. Digitalism entails that right now you can truly say: "I will be one greater extended body after another."

114. Genes for deep fractal organs

An extended human, like every kind of human, begins with the fusion of an extended sperm and egg. It starts as an extended zygote, which encodes its genotype. The zygote divides to make two cells; each of those cells divides to make four cells; and so it goes. Every division of an extended cell is an *intrinsic doubling*. The intrinsic doubling of a cell goes like this: (1) the parent cell divides into two offspring; (2) each offspring is twice as small; (3) each offspring runs twice as rapidly and efficiently; and (4) each offspring runs twice as reliably (with half as much damage during metabolic activity).

As cells double intrinsically, cell networks double intrinsically. The previous cell network turns into one with twice as many intrinsically doubled cells and at least twice as many internal connections. It doubles in density. As cell networks double intrinsically, they form ever deeper fractal organs. Many earthly human organs are shallow fractal structures produced by just a few iterations of a self-repeating growth pattern. They have finitely deep fractal structure (Goldberger et al., 1990; Bassingthwaighte et al., 1994). Fractal patterns occur in nerve cells; in the airways in the lungs; and in the intestines. Humans are animate fractals with a few levels of fractal depth. The earthly human genotype specifies the earthly fractal patterns in the human body.

The intrinsic doubling of all the cells in an organ takes the organ as a whole to the next level of fractal refinement. For example, when all the cells in a lung double intrinsically, the branching bush of airways in that lung becomes twice as large and precise. The number and functional power of these airways doubles. An n-th level organ $O(n)$ has gone through n iterations of its growth pattern. Every n-th level extended body $B(n)$ is composed of n-th level organs. All its organs (like its limbs, sense organs, and nervous system) have gone through n fractal refinements. The fractal refinement of an organ increases its depth by one. Likewise the fractal refinement of a body increases its depth by one. Every next extended body $B(n+1)$ is composed of organs of the form $O(n+1)$.

An extended body $B(n)$ starts with an extended zygote $Z(n)$, which carries an n-th level extended genotype. Any n-th level extended genotype encodes a growth plan which drives every organ, and therefore the entire body, through n iterations of its fractal pattern. This recursive growth plan is part of the n-th level extended soul. Any n-th level soul contains all the if-then rules for the entire life of the body $B(n)$. It is the n-th level body-program. But every n-th level soul $S(n)$ is surpassed by an intrinsically doubled soul $S(n+1)$. All the extensions of any soul are in its essence.[6] The essence of any human is closed under optimization, idealization, and extension. Of course, all the organs of an extended human body must still fit within the essential human body-plan.

115. The way of all flesh

An extended human has extended human organs. From each extended body to its successor, the functionalities of these organs double. For any physiological function F, if any extended body can do F, then the next version of that extended body can do F with twice the speed, precision, efficiency, and reliability. For digitalists, the way of all flesh is the endless doubling of the powers of all organ systems. Every extended human body has the organ systems found in earthly, optimized, and idealized humans. These include the skeletal, epidermal, sensory, computational, metabolic, repair, and reproductive systems. There is no need to discuss all these extended organ systems in detail. Here we focus on the extended skeletal, computational, and sensory systems.

The Skeletal System. An earthly human looks like a shallow tree (a bush). Limbs branch out from limbs: (1) there is a central trunk; (2) two arms branch off at the top and two legs branch off at the bottom; (3) five fingers branch from each arm and five toes from each leg. So an earthly body is a *bush* with three levels of branching (the fractal depth of its skeleton is three). Further iterations proceed by putting fingers on fingers and toes on toes (see Moravec, 1988: 102–8; 2000: 150–4). However, the fractal refinement of the earthly body is irregular. It branches into two limbs (arms or legs) and then into five sublimbs (fingers or toes). But extended bodies follow regular patterns.

The regular branching of extended skeletal systems follows the powers of two: the trunk grows two branches; each branch grows four sub-branches; each sub-branch grows eight sub-sub-branches; and then sixteen; and so it goes. Hence the n-th level skeleton is defined by iteration from zero to n. The initial limb $L(0)$ is the trunk. It is a thick stalk which contains the internal organs of the body. For any m less than n, the successor limb $L(m+1)$ is $L(m)$ plus 2^{m+1} sticks on the end of each stick in $L(m)$. Each successor stick is twice as short, thin, fast, efficient, reliable, dexterous, and precise as the one which supports it. The growth of the skeleton stops when m reaches n. The result is the adult n-th level extended skeleton $L(n)$. Each adult skeleton $L(n)$ has the form of a branching tree with n levels. Figure VIII.1 partly illustrates the fractal refinement of the skeleton. It shows the first four iterations; but it cannot show all the fingertips, and it does not show the head.

At each iteration, the length of each new branch is half as long and half as thick as that of its previous branch. Yet all branches remain equally strong. The strength of each branch is derived from the intrinsic doubling of muscle and bone cells. Each muscle fiber splits into two muscle fibers that are twice as strong, run at twice the speed, and are twice as efficient (the same energy produces twice as much force). The endurance of the fibers doubles. The doubling of muscle fibers doubles the acuity (the fine motor control) of the branches. Both legs and arms have intrinsic doublings. At each iteration, the number and fineness

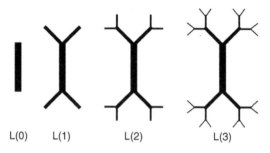

L(0)   L(1)   L(2)   L(3)

Figure VIII.1 The first four extended body structures.

of fingertips and toetips doubles. Hence the number of neural paths needed to control the limb doubles. An intrinsically doubled limb needs an intrinsically doubled brain to control it. Extended skeletomuscular systems require extended brains.

The Sensory System. At a high level of functional abstraction, an eye is a hollow ball with a lens at one end and a sensor at the other. The *n*-th level extended eye E(*n*) starts with an initial form E(0). It grows through all its immature forms. These are the eyes E(*m*) for *m* less than *n*. It finally converges to its adult form, the eye E(*n*).

The initial eye consists of an initial lens and retina. The initial lens F(0) is just a transparent disk – it does not focus. The initial retina R(0) is just a single simple photocell. It detects only the difference between light and dark. But every eye E(*m*) is surpassed by a successor eye E(*m*+1) with twice as much visual power. The lens F(*m*+1) is twice as powerful as the previous lens F(*m*). Each next lens has twice the focal length; twice the zoom speed and focusing precision; twice the angle of view; twice the aperture control; twice the optical precision. The retina R(*m*+1) is twice as powerful as R(*m*). Each successor eye has twice the resolving power (visual acuity) of its predecessor.

Any retina is a grid of photosensitive nerve cells (rods and cones). It can be extended by intrinsic doubling. The next iteration of the retina is made by (1) dividing each cell in half on each dimension (vertical and horizontal) to make four times as many cells, each of which is twice as small on each dimension; (2) making every cell run twice as fast, twice as efficiently, and twice as reliably. The intrinsic doubling of the retina increases the number and variety of photoreceptor cells. Figure VIII.2 shows the first few iterations of the retina. Each intrinsic doubling of the retina also intrinsically multiplies the wiring behind the sensor cells and the number of fibers in the optic nerve.

Each iteration of the eye has greater visual functionality. It is sensitive to a greater part of the radiant spectrum – it sees more colors. It sees further into the extended spectrum (into its analogs of infrared and ultraviolet, and even further). It has higher resolving power and

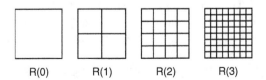

Figure VIII.2   A series of extended retinas.

magnification. It can function like a camera with wide angle, macro, and telephoto lenses. It can function like a microscope and a telescope. It can switch between these configurations with increasing rapidity and precision. It can focus ever more rapidly and accurately, with ever more extreme control over depth of field. It is ever more sensitive to lower levels of radiation. It can apply a variety of filters to its input: it can polarize its input, or selectively filter in or out certain parts of the spectrum.

The Computational System. At a high level of functional abstraction, an earthly brain is a network of neurons. It is a connect-the-dots network of cellular computers (Section 23), which themselves are connect-the-dots networks of molecular computers (Section 19). The n-th level extended brain $N(n)$ starts with an initial form $N(0)$. It grows through all its immature brains, which are the $N(m)$ for m less than n. It finally converges to the adult brain $N(n)$. Of course, every adult brain $N(n)$ is itself surpassed by the extended brain $N(n+1)$. These are the brains of the next extended human bodies, living in the next extended universes, running on the next extended gods.

The initial brain is an idealized human brain. Each successor brain develops by intrinsic doubling. It is derived from its predecessor brain by doubling its number of neurons and increasing their synaptic connections. Each successor neuron is twice as small, fast, efficient, precise, powerful, and reliable as its predecessor – it is twice as intelligent. It is a computer with twice the internal memory and processing speed. Nerve cells are fractals taken to finite depths. An earthly nerve cell is a branching structure of axons and dendrites. The depth of branching is only a few levels. Each further iteration of nerve cell growth implies an extra level of branching—the existing branches sprout finer branches. Neural networks are fractally extended in extended humans.

An earthly human brain can perform a wide range of computations. Intrinsic doubling extends that computational range. These extensions overcome the bottlenecks in human information processing. There are three well-known bottlenecks in the earthly brain: (1) the time lags in visual processing; (2) the small number of items that can be stored in visual working memory; and (3) the inability to multitask (Marois & Ivanoff, 2005). These constraints are steadily lifted by extension. From each extended brain to its successor, the time lags are cut in half; the number of items in visual working memory doubles; the number of tasks that can be done simultaneously doubles.

Extended bodies are saturated with intelligence. While any unextended body has only one brain, any extended body has many brains. It has level after level of sub-brains that perform subroutines. While strategic or executive decisions are made in the brain in the head, tactical decisions are delegated to the sub-brains. Each limb contains a sub-brain at each joint. The nervous system is distributed throughout the limbs. This distribution makes the neural processing speed of the entire extended body much faster.

Systems of organs are collectively extended. Consider an extended eye–brain–arm circuit. The circuit is iterated by iterating the eye, the brain, and the arm. The series of iterations of the eye–brain–arm circuit converges in the limit to an infinitary eye–brain–arm circuit. Each iteration produces functional doubling of the retina; the visual cortex; the maps from the visual cortex into the motor cortex; the motor cortex itself; and the arm. A good test for the functional excellence of such a system is the ordinary game of darts. At each iteration, the eye–brain–arm circuit can throw a dart twice as far, twice as fast, and twice as accurately as the previous iteration. After a few iterations, an extended human animal can throw a small dart as fast, as forcefully, and as accurately as an earthly human can shoot a bullet from a high-powered rifle with an optical targeting scope.

The Other Organ Systems. An extended body has extended versions of all earthly human organs. All these organs grow from initial forms, through many intrinsic doublings, to their mature forms. And they are all surpassed by the greater organs of the next extended bodies in the next universes running on their greater gods. While there is no need to go into all these organs in detail, it will be helpful to say a few words about some of them.

An *extended metabolic system* can dynamically reconfigure itself to use any energy source in its universe. It is a physically universal energy system. The cell-networks that are specialized to absorb and process energy are distributed throughout the body. They are not concentrated in the first-level trunk. Extended reproductive systems are innately hermaphroditic – they can switch by training between male and female. Of course, all extended human children still have one male and one female parent.

As the result of their own sturdiness and their *extended repair systems*, extended humans live longer and longer lives. The life of any first-generation extended human H(0) is as long as that of an idealized human. But the life of each next-generation extended human H(n+1) is twice as long as that of any human of type H(n). Extended humans live for thousands, millions, billions, and trillions of years. For any finite length of time, there are some extended human lives that are longer than that time. Nevertheless, every particular extended human life is only finitely long. The length of any extended human life is surpassed by some finite number, which it cannot exceed or overcome. Every extended human dies. And it lives in an ecosystem which will also die, in a universe in which all structure will vanish, and whose underlying god will cease to operate. Every god in every progression of extended gods operates only finitely. Fortunately, every progression of finite gods is surpassed by some infinite gods. But those infinite gods make infinite universes, which contain infinite ecosystems, infinite lives, and infinite bodies.

IX
Infinite Bodies

116. Subtle physics

Our local god lies within our local progression of ever greater gods. And since every god supports some universe, our local progression of gods supports our local progression of universes. Obviously, our universe lies within our local progression of universes. And, since digitalism entails that every progression is surpassed by many infinite limits, our local progression of gods is surpassed by many infinite *limit gods* and our local progression of universes is surpassed by many infinite *limit universes*. These limit gods and their limit universes are all infinitely intrinsically valuable and complex. An infinitely complex thing is *infinitary*. Alternatively, it is *subtle*. It will be helpful to focus on one subtle god and its subtle universe. This god is $G(\omega)$ and its universe is $U(\omega)$.

Any subtle universe has continuous space and time. It is filled with infinitely many particles of *excelsium*. Any particle of excelsium is an infinitary computer with some finite size.[1] It has infinite memory and can perform *supertasks*.[2] A supertask is an infinite series of tasks done in some finite volume of space-time. A supertask done in some unit of time (such as one minute) is a progression of finitely complex (*finitary*) tasks done by the *Zeno fractions* in that unit of time. These are the fractions 1/2, 3/4, 7/8, 15/16, and so on. The first task is done by time 1/2; the second by 3/4; the third by 7/8; and so on. At time one, infinitely many tasks have been done, and the supertask is complete. More precisely, any supertask performs ω-many tasks in one unit of time. Any particle of excelsium can form bonds with infinitely many other particles of excelsium.

Any subtle universe begins with some initial excelsium particles, which recursively self-improve: they create superior versions of

197

themselves in the next moment. And they form increasingly complex networks, whose particles are bound by lines of force. Subtle evolution proceeds by *acceleration*. A process accelerates if and only if its next phase always runs twice as fast as its previous phase. Supertasks proceed by acceleration. Thus each generation of more complex structures appears twice as fast as its previous generation. After some finite time, networks appear which contain infinitely many excelsium particles linked with infinitely many bonds. These particles run infinitary programs.

Subtle chemical and biological structures appear quickly in subtle universes. These subtle biological structures are subtle organisms living in subtle ecosystems. Subtle ecosystems are infinitely rich and diverse. After some finite time, subtle evolution produces subtle hominids and then subtle humans. Since our local progression includes our earthly universe, all earthly humans have subtle future counterparts in $U(\omega)$. You and I have subtle future counterparts in $U(\omega)$, as do Deucalion and Pyrrha. According to digitalism, right now you can truly say: "I will be a subtle body."

117. Subtle physiology

A subtle human body, like every possible human body, begins with the fusion of a sperm and an egg. It starts as a subtle zygote. The subtle zygote is a particle of excelsium, about the size of an earthly human zygote. It encodes the subtle genotype. Any subtle genotype is a part of a subtle soul, and any subtle soul is the limit of a progression of ever more excellent souls.[3] Your earthly soul lies in a progression which has many limits; every one of those limits is one of your subtle souls. But every limit of any progression of souls is in the essence of that soul. So your subtle souls are in your essence.

Any subtle soul contains a subtle growth plan. Following this plan, the subtle zygote divides; its offspring divide again; and so it goes. Its descendent cells are all particles of excelsium. Every division of a subtle cell is an *intrinsic doubling* (Section 114). But these intrinsic doublings accelerate: each next division happens twice as fast. After one unit of time, infinitely many divisions have taken place. As these cells divide, some of them become ever smaller, while others retain their finite sizes. As these cells divide, they form connect-the-dots networks in which the dots are cells and the connections are lines of force. They fill out the body-plan of the subtle body, which includes all and only the organs in the human body. Subtle humans look much like earthly humans.

A subtle body has the organ systems found in earthly, optimized, idealized, and extended bodies. These systems are skeletal, epidermal, sensory, computational, metabolic, repair, and reproductive. All subtle organs are the limits of progressions of extended organs. Any subtle system of limbs is the limit of a progression of extended systems of limbs; any subtle eye is the limit of a progression of extended eyes; any subtle brain is the limit of a progression of extended brains. Progressions of these organs were defined in Section 115. More formally, any subtle organ $O(\omega)$ is the limit of some progression of extended organs of the form $\langle O(0), O(1), O(2), \ldots \rangle$. Any subtle body $B(\omega)$ is the limit of a progression of extended bodies. It is composed entirely of subtle organs.

Every subtle organ can perform all its functions with infinite excellence. It can perform supertasks. A subtle eye can perform *visual supertasks;* a subtle brain can perform *cognitive supertasks;* a subtle system of limbs can perform *motor supertasks.* And the subtle body as a whole can perform *behavioral supertasks.* It can perform artistic and athletic supertasks. Some of these supertasks are solitary while others are social. The best way to introduce them is the way of example – but the examples given here are merely schematic. For details, see Steinhart (2002, 2003, 2007b).

118. Writing an infinite book

A *Zeno tape* can be used to define a simple supertask. A Zeno tape is a finite length of paper divided into infinitely many cells. It is a line segment running from zero to one. The first cell starts at zero and ends at 1/2; the second starts at 1/2 and ends at 3/4; the third goes from 3/4 to 7/8; and so it goes. Each next cell is half as wide as its predecessor. The tape has a cell ω at its rightmost end. This limit cell is exactly as wide as one point. Figure IX.1 partly illustrates a Zeno tape. Only the first six cells are shown.

For the sake of illustration, say that *Helen* performs supertasks. Helen is a subtle body. Her subtle brain presently stores in its memory an infinitely long string of binary digits (bits).[4] She wants to write this infinite series down on her Zeno tape. Suppose the string starts like

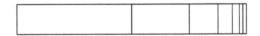

Figure IX.1 Partial illustration of a Zeno tape.

this: 1100. Helen uses her pen to write down the first digit 1 in the leftmost cell of her tape (cell zero) in 1/2 second. She writes the next digit 1 in cell one in the next 1/4 second. She writes 0 in cell two in the next 1/8 second. Then 0 in cell three in the next 1/8 second. Then 0 in cell four in 1/16 seconds. And so it goes.

As she moves to the right, the cells get smaller and smaller. Her hand and pen also get smaller and smaller. Writing each next number in a next space that is always half as big as before is an example of *Zeno compression*. Her hand and pen become compressed. As Helen goes on, she has to write faster and faster. Writing each next digit in a time that is always twice as small as before is an example of *Zeno acceleration*. Since Helen is accelerating, she finishes her supertask in one second. At the limit time one, her pen is at the limit cell ω. She has written every finite number on her tape.

Since continuous time is infinitely divisible, and since subtle bodies act in continuous time, they can perform *supertasks of supertasks*.[5] Supertasks of supertasks can be nested infinitely deeply. One supertask of supertasks consists of generating an infinitely long list of infinitely long bit strings. By performing an elaborate mental computation inside her brain in one second, Helen generates an infinitely long list of infinite bit strings. Perhaps she was simulating a universe, and her computation involved figuring out all the ways that it can be improved. Each bit string describes one of these ways.

After performing her computation, the brain of Helen contains an infinite data structure in its memory. But now she wants to write it down so she can pass it on. She will make a book with infinitely many pages. She will make a *Borges Book* (1964: 58). Each page of this book is a Zeno tape. The first page is 1/2 inch thick; the next page is 1/4 inch thick; and so on. By compression, infinitely many pages fit into the Borges Book. By accelerating in the interval from zero to 1/2, Helen writes her first infinite bit string on the first page in 1/2 second; by accelerating in the interval from 1/2 to 3/4, she writes the next bit string on the second page in 1/4 second; and so it goes. At one second, she has finished her supertask of supertasks. She has written her Borges Book.[6]

119. Eating an infinite feast

After writing out the Borges Book, Helen is hungry. Any organism has to eat, and subtle bodies often eat infinite feasts. Helen will eat all the treats in her infinitely complex yet finitely sized lunch box. This lunch box is divided into compartments as shown in Figure IX.2. Each next

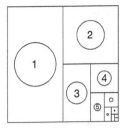

Figure IX.2   The box of infinite treats.

compartment is half the size of the previous compartment. Each compartment holds a treat, illustrated by the numbered circles in Figure IX.2. Each next treat is half the size of the previous treat, so the amount of food in the lunch box is finite.

Helen is a hungry subtle body; to satisfy her hunger, she will eat every treat in her lunch box in one minute. She accelerates. Within the first 1/2 minute, she reaches into the first compartment, grasps the first treat (the treat labeled 1) with her fingers, brings it to her mouth, and devours it. Thus she eats the first treat in 1/2 minute; she eats the next treat in the next compartment in 1/4 minute; by always doubling her speed, she eats every treat in the box in one minute. At one minute, the box is empty and she is satisfied.

Of course, since these treats grow ever smaller, and lie in ever tinier compartments, eating them requires ever greater dexterity. The hand of Helen is a superbush; it is an endlessly ramified tree of fingers on fingers. It is the limit of a progression of extended hands (see Section 115). For any finite n, Helen can open her hand at the n-th joint. When she opens it at the first joint, her hand has two fingers, each of thickness 1/2. When she opens it at the second joint, her hand has four fingers, each of thickness 1/4. When she opens it at the n-th joint, it has 2^n fingers of thickness $1/2^n$. To get the n-th treat out of its compartment, she opens her hand to the n-th joint.

Helen enjoys these treats. They all have different flavors, which she can taste. She appreciates the skill of the subtle cook. But cooking is just one type of art, and subtle bodies make subtle art of every kind. Any subtle body can paint a picture using a palette with infinitely many pigments (corresponding to the infinitely many colors of subtle light). Hence subtle bodies can paint and see infinitely beautiful pictures. The beauty of any such picture exceeds every finite degree of beauty. Since subtle sound has infinitely many tones, subtle bodies can compose and hear infinitely beautiful pieces of music. Their sculptures and architectural works are likewise infinitely beautiful.

120. Seeing an infinite picture

A body with super-perception can perceive infinite detail in finite time. It does this by performing a perceptual supertask. A subtle body can perceive infinitary fractals (it can see, in one glance, all the infinite detail of any Koch curve or fractal plant). Super-perception is nicely illustrated by the *Royce Map* (Royce, 1899: 506–7).

The first Royce Map $M(0)$ is just a square with a cross inside it. Each next Royce Map $M(n+1)$ is produced by copying the previous map $M(n)$ and drawing a cross in its lower right quadrant. Figure IX.3 shows the first four iterations of the Royce Map. Drawing the Royce Map is a supertask. When this task is completed, the limit Royce Map $M(\omega)$ is infinitely self-nested. The limit map contains an exact copy of itself (which contains an exact copy of itself, which contains an exact copy of itself, which...). Here the Royce Map itself is just the limit map. The Royce Map is an infinitely rich fractal.

After eating her infinitary lunch, Helen draws the Royce Map on a sheet of paper. She hands it to her daughter, Hermione. Like her mother, Hermione is a subtle body. Thanks to her subtle eyes and brain, she has the power of super-perception. She can successively perceive the members of an infinite series of increasingly rich visible objects and her perceptions converge in the limit to the limit object of that series.

Hermione will *see* all the detail of the Royce Map in one second. To do this, she starts with the initial Royce Map; she then zooms in on each next level. Her eyes are like microscopes that can always focus more precisely on ever smaller details. As she accelerates, she zooms on the extra detail in each lower right quadrant. She sees $M(1)$ in 1/2 second; she sees $M(2)$ in 3/4 seconds; and so it goes. At one second, she has visually focused on all the detail in the entire map. Her infinite visual memory contains an exact image of $M(\omega)$. Hermione is ready for her next task – what will it be?

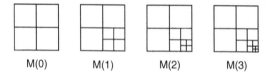

M(0)   M(1)   M(2)   M(3)

Figure IX.3 The first four iterations of the Royce Map.

Helen hands Hermione the Borges Book and tells her to focus on a specific page. Hermione grasps immediately that the infinite bit string describes a universe. Of course, she also grasps that it also contains some instructions: bring this universe into existence by running it in your brain; after generating it, figure out all the ways to improve it; produce as many offspring as there are ways; hand one way down to each offspring along with these instructions. Since she is good, Hermione carries out these instructions.

121. Thinking infinite thoughts

A subtle body can create, manipulate, and perceive infinitely detailed fractals (Mandelbrot, 1978; Prusinkiewicz & Lindenmayer, 1990). Since they are infinitary things, such fractals are the natural objects of perception, thought, and action for subtle bodies. For example, a subtle body can draw Koch curves and islands; it can make dragon curves; it can paint a fully detailed picture of the Mandelbrot set. It can build Sierpinski gaskets and Menger cubes. It can make the Lakes of Wada, which involves digging an infinitely detailed fractal system of canals on an island (Koetsier & Allis, 1997: 293).

Of course, fractals are not the only infinitary objects that subtle bodies can create, manipulate, and perceive. There are many other types of infinitary objects. Subtle bodies can also work with infinitary connect-the-dots networks. For example, they can make models of the iterative hierarchy of pure sets up to any finite rank (Hamilton, 1982: chs. 4 & 6). The zeroth rank of this hierarchy contains the empty set. The next rank contains every combination of sets from all previous ranks. Sets on lower ranks are members of sets on higher ranks. When a subtle body makes a model of the hierarchy, the sets are dots and the connections are instances of the membership relation.

A subtle body can perform infinite computations on the mental representations in its super-memory. Some of these infinite computations are *super-proofs*, which involve infinitely many steps. Many of these super-proofs find solutions by exhaustive search. For example, consider Goldbach's Conjecture, which states that every even number greater than two is the sum of two primes. A subtle body can determine whether that conjecture is true or false by checking every even number greater than two. It tests 4 in 1/2 second; it tests 6 in 1/4 second; it tests 8 in 1/8 second; and so it goes. If any even number fails to be the sum of two primes, it knows that Goldbach's Conjecture is false. After one second, it has examined every even number. If it has not found any counterexample by one second, then there are none, and it knows that Goldbach's Conjecture is true.

A subtle body can do anything that can be done by any finitary machine.[7] It can mentally perform all possible finite computations. So it can exactly simulate any finitely complex universe. Its simulation is not an approximation – it is an exact reproduction. So a subtle body can contain a copy of any finitary universe. If there are lives in some finitary universe, the subtle body can exactly replicate them (Moravec, 1988: 178–9). The copy of a universe in a subtle body is as real as the original. Any subtle body can simulate every possible finitary universe. And for every finite god g, there exists some subtle body b such that b can exactly simulate g; by simulating g, it simulates the universe that supervenes on g, including all the bodies therein. Of course, these simulations can be nested: subtle bodies may simulate great gods, whose universes contain great bodies; those simulated great bodies may in turn simulate lesser gods, whose universes contain lesser bodies. This nesting of simulations permits *promotion* (see Chapter V). A body that simulates a universe may decide to promote some of the simulated bodies therein. Some subtle bodies may simulate an infinite hierarchy of nested simulations. If such subtle bodies exist, then they resemble the Deity described in Sections 60–66. But any such subtle bodies have evolved from simpler antecedents, all the way back to the initial god Alpha.

122. Making infinite love

Subtle bodies interact by playing infinitary games. A good introduction to such games is an infinitely long *driving game*. It is a variant of the classical infinite digits game (Hamilton, 1982: 189; Neeman, 2004). Deucalion and Pyrrha are the two players – and this game defines their intimacy. Once upon a time, their predecessors were finite (Sections 92–115); but they have made infinite progress. Deucalion and Pyrrha are each the limits of infinitely long sequences of increasingly excellent lives.

Deucalion and Pyrrha are both driving down a road in the same car. This car is like an ordinary car but with an important difference: it has two steering wheels. Hence both of its two front seats are driver's seats. They drive 1/2 mile when they come to a fork in the road. Deucalion moves first. At the fork, he decides whether to go left or right. After Deucalion makes his choice, our two players drive another 1/4 mile and they come to another fork in the road. At this fork, it's Pyrrha's turn to decide whether to go left or right. After Pyrrha decides, they drive

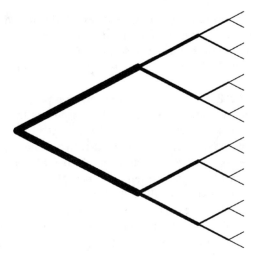

Figure IX.4   The first four levels of the endlessly forking roads.

another 1/8 mile and come to another fork. Once more, it's Deucalion's turn to decide which way to go (left or right). The series of forks is infinite. It's always half the distance to the next fork. And, at each fork, the road shrinks to half its width and the car (and its drivers) also shrinks. A few levels of the branching tree of endlessly forking roads (each half as long and thin) are shown in Figure IX.4.

Once Deucalion and Pyrrha have gone one mile, they have driven through infinitely many forks and made infinitely many turns. If they drive at a constant speed, say one mile per hour, then it only takes them one hour to go through infinitely many forks. Of course, they don't go through all the forks at the same rate. Since the first fork is 1/2 mile from the start, it takes 1/2 hour to get to it. Since the next fork is only 1/4 mile from the first fork, it only takes 1/4 hour to get to it. And it takes only half the time to get to the next fork. Deucalion and Pyrrha are doing Zeno acceleration through the forks. So, in one hour, they've gone through them all. They have reached a point on the finish line.

At each fork, Deucalion and Pyrrha use their strategies to determine which way to go. A strategy is a list of rules. The simplest strategy just has one rule. Here are four different one-rule strategies: always turn left; always turn right; always do what the other driver did; always do the opposite of what the other driver did. More complex strategies have longer lists of rules. The most complex strategies have a rule for every fork in the road. Since each *fork* is reached by a finitely long series of turns (and every finitely long series reaches a fork), the most complex strategies have a rule for every finitely long series of turns. Just as there are infinitely many finite numbers, so there are infinitely many finitely long lists of turns. The most complex strategy involves an infinitely long list of rules.

Deucalion and Pyrrha each use infinitely complex strategies, composed of infinitely many if-then rules. Each rule has this form: if the other driver is making turn x and my list of past turns is y, then I'll append x to my list of turns and my next turn will be z. For example, one of the rules in Deucalion's strategy might be: if Pyrrha is turning L and my list of past turns is LR, then I'll append L to my list of turns to make LRL and my next turn will be R. Of course, since these drivers are subtle bodies, they have each memorized their infinite strategies, which they store in infinite mental rule books. And, as they drive, they memorize the turns. Each driver has an exact mental list of past turns, which it uses to mark its place in its mental rule book. More technically, each driver is an *infinite state machine*. Its set of inputs is {L, R}. Its set of outputs is also {L, R}. Its set of states is the set of all finitely long series of turns. Its rules encode its transition function and its production function. Its transition function maps each (turn, list of turns) pair onto the next list by appending the turn to the list. Its production function maps each (turn, list of turns) pair onto the turn it will make when it comes to the next fork.

At every fork, each driver has to update its memory and apply a rule. After applying their rules infinitely often, Deucalion and Pyrrha reach an end point in their garden of forking paths. This point is determined by the interactions of their strategies. It is painted either gray, pink, blue, or golden. If it's gray, then they both lose the game; if it's pink, then Pyrrha wins and Deucalion loses; if it's blue, then Deucalion wins and Pyrrha loses; if it's golden, then they both win. The utility of any point is its number of winners. The golden points are the best points – any journey to any golden point is a *journey of love*. It is a journey in which their infinite strategies are perfectly *harmonized*. Deucalion and Pyrrha strive to keep their strategies harmonized – they strive to win together. From the finitary Deucalion and Pyrrha to their infinite selves, there has been an infinite chain of ever more loving couples, ever more intensely harmonized, ever more intensely intimate.

123. Playing infinite games

As an illustration of an infinitary game for super-athletes, consider infinitary basketball. It is a generalization of ordinary finitary basketball.

Finitary basketball is played in a finite volume of space-time (a court). A court has two sides, each of which can be thought of as a unit cube. Finitary basketball involves two teams with many players each. These players handle a ball, bouncing it, passing it to and fro, and so on. Each team has a goal which is a basket. A team gains points when one of its players throws the ball through the basket of the other team. Of course, the rules involve many details – but some familiarity with finitary basketball is assumed. It won't be discussed any further.

The definition of infinitary basketball starts at level one. At level one, it is just like finitary basketball. From level one to level two, the difference is just this: the ball and the baskets each shrink by half. The players remain the same size, but they can move twice as fast and precisely. At level two, they need greater hand–eye coordination to handle this smaller ball. They need to be able to pass it back and forth and to bounce it up and down. They also need to be able to throw this smaller ball through the smaller baskets. Basketball at level three is defined analogously: the ball and the baskets shrink by half again. Again the players remain the same size, but can move twice as fast and precisely. Iteration defines basketball at level four, level five, and so on to level ω. At level ω, the players are playing with a ball that is the size of a point and trying to shoot it through a basket that is also the size of a point (that is, they're just trying to hit the basket).

Infinitary basketball is played by moving from level n to level $n+1$ each time control of the ball changes (either when a goal is scored or when one team steals the ball). After some finite period of time, the team with the highest score wins. Equal scores (whether finite or infinite) mean a tie. Playing infinitary basketball requires a body with infinitely precise perception, cognition, and volition. An infinitary basketball player needs eyes with infinitary resolution. They need a brain that can compute the trajectories of arbitrarily small and arbitrarily fast objects. They need a brain that can accelerate. They need a motor system that can move in infinitely precise ways. Their muscles also need to accelerate. They require infinitely fine hand–eye coordination. An infinitary basketball player needs an infinitary body (like the limit of some progression of extended bodies). Analogous techniques define infinitely precise versions of games like handball, soccer, and pool.

124. Subtle social bodies

Subtle bodies interact by playing infinite games. Their social and political lives are tournaments of multiplayer infinite games. Many infinite games are mathematical devices for the construction of infinite series of situations.[8] Moves in these infinite games typically involve selecting numbers or constructing sets. For subtle bodies, these actions are infinitely psychologically rich. Emotions run high in these games. For subtle bodies, which can perceive infinite detail, real numbers are infinitely beautiful or ugly. The selection of an infinitely beautiful number is an infinitely pleasurable act; the selection of an infinitely ugly number is an infinitely painful act. By playing these infinite games, subtle bodies eat, drink, walk, run, dance, compete, cooperate, make love, marry, and bear and raise children. They engage in infinitely intensified versions of all earthly human actions. And they engage in actions which are so intense that we cannot comprehend them.

After maturity, any subtle human lives for some countably infinite number of discrete units of time. They retain their full functionality for countably infinitely many discrete moments. They thus live forever in a weak (countable) sense. Every subtle human body is a part of an infinitely long life – it is *immortal*. Paradoxically, this immortality does not preclude death. Subtle bodies die. To say that a subtle body dies means exactly that its life comes to an end. Subtle lives end because there are numbers greater than any countable numbers; hence there are lives longer than any countably long life. Every countably long life is surpassed by uncountably long lives. And, relative to those uncountable lives, merely countable lives look extremely short indeed. Countable lives are immortal relative to finite lives; yet relative to uncountable lives, they are tragically mortal. For every number on the Long Line, there is a corresponding degree of mortality. Subtle lives have ω-mortality. Fortunately, every subtle life is revised – its life is improved into some greater life.

125. Mechanical archons

The least infinite number ω is only countably infinite. Another term for ω is $\aleph_0$, and this $\aleph$-notation will be used in this section. As discussed in Section 75, every infinity $\aleph_n$ is surpassed by a greater infinity $\aleph_{n+1}$. Hence $\aleph_0$ is surpassed by $\aleph_1$. All the infinities above $\aleph_0$ are uncountable. The rules for the divine hierarchy (Section 77) entail that our local progression runs into the uncountable. One of the descendents of our local god is an uncountable god $G(\aleph_1)$; but $G(\aleph_1)$ has a descendent $G(\aleph_2)$; and so it goes. These gods support uncountably complex universes $U(\aleph_1)$, then $U(\aleph_2)$, and so on.

Uncountably complex universes contain uncountably complex computers and networks of computers. They contain uncountably complex cells, organs, and bodies. Since our universe has descendents which are uncountably complex, you and I have uncountably complex future counterparts in those universes (as do Deucalion and Pyrrha). We have counterparts in $U(\aleph_1)$, $U(\aleph_2)$, and so on. Hence right now you can truly say: "I will be an uncountable body". An uncountable body can do anything that a countable body can do. It can perform any supertask. But an uncountable body can do more: it can perform *hypertasks*. A hypertask compresses uncountably many operations into some finite volume of space-time. Any body that can perform hypertasks is an *archon*.

An archon works in a very rich infinitary universe. After all, hypertasks cannot be done in ordinary continuous time (Clark & Read, 1984). A richer kind of time is needed (Barrett & Aitken, 2010). Kitcher talks about a kind of temporal medium that is far richer than time (1984: 146–7). One way to define *hypertime* is to use the *hyperreal number line* (Robinson, 1996) or the *surreal number line* (Conway, 2001). Both of these lines have infinitesimals. However, describing these lines is extremely technical. All we need to know is that hypertime is a temporality rich enough for any hypertask.

An archon needs to work in some space. Perhaps *hyperspace* is some multidimensional manifold of hyperdimensions (such as a multidimensional coordinate space in which each coordinate axis is the hyperreal or surreal number line). Archons live in universes with hyperphysics. Although hyperphysics is intriguing, it's more interesting to focus on the archons themselves. They have *hyperperception*, *hypercogniton*, and *hypervolition*. Archons are interested mainly in other archons. Finite or merely countably infinite bodies are far too weak for them. Space for an archon is a social network of other archons. Archonic interactions are uncountably complex games.

All archons are based on hypercomputers – on extensions of Turing machines and register machines to higher infinities. Since there are many types of hypercomputers, there are many types of archons. Only five types are mentioned here. Since much advanced mathematics is needed to describe these archons, their descriptions are omitted. The first type includes all *infinite time Turing minds*. These are intelligent infinite time Turing machines (see Hamkins & Lewis, 2000; Hamkins, 2002). The second type includes all *α-Turing minds*. These are intelligent α-Turing machines (see Koepke, 2007). The third type includes all *infinite time register minds*. These are intelligent infinite time register machines (see Koepke, 2006a). The fourth type includes all *ordinal Turing minds*. These are intelligent ordinal Turing machines (see Koepke, 2005, 2006b). The fifth type includes all *ordinal register minds*. These are intelligent ordinal register machines (see Koepke & Siders, 2008). All these archons are uncountably infinitely powerful.

Archons can create, manipulate, and perceive structures of uncountably transcendental beauty. The beauty of any such structure exceeds every finite and countable degree of beauty. Archons use their uncountably subtle bodies to perform physical hypertasks. For example, starting with one solid 3D ball in its hands, an archon uses its infinitely sharp fingertips to cut that ball into finitely many pieces; it then rearranges those pieces, without bending or distorting them, into two solid balls, each of which is the same size as the original. By transforming one solid ball into two copies of itself, the archon illustrates the Banach–Tarski theorem (Hamilton, 1982: 186).

Any archon can produce uncountably complex structures. These are typically models of sophisticated mathematical theories. For any cardinal k, there is some archon that can build a model of the hierarchy of pure sets up to rank k. There are archons that can build models of ZFC set theory. The model is a connect-the-dots network in which the dots are sets and the connections are instances of the membership relation. If an archon builds a model of ZFC, then it can empirically test every open question of ZFC with respect to that model. The body of that archon (as well as its environment) is at least as complex as some level of the set-theoretic hierarchy whose index is an inaccessible cardinal (Hamilton, 1982: 230–3). An archon that can build a model of ZFC is an *inaccessible archon*. But these archons are surpassed by even greater archons, such as the *measurable archons*, the *supercompact archons*, and so on.

Any archon lives for some length that is proportional to its complexity. An inaccessible archonic life consists of an inaccessibly long series of archonic bodies. And yet, no matter how long any archon lives, there is some number that is greater than the length of its life. Hence every archon dies. Archons die because, however long they live, there are lives that are longer. To die is just to be surpassed by some greater life. Of course, every archon that dies is reborn – its life is revised into some greater life. But there are no top-level archons. Every archon is surpassed by a proper class of greater archons.

X
Nature

126. Abstractness

On the basis of the Indispensability Argument (Section 33), digitalists affirm the existence of an abstract background of mathematical objects. For digitalists, the most prevalent abstract objects are sets. Among sets, the simplest is the *empty set,* which does not contain any members. The empty set is denoted {}. But simpler sets are members of more complex sets. Hence the empty set is a member of the set of the empty set, which is written as {{}}. And both {} and {{}} are members of {{}, {{}}}.

Since simpler sets enter into the constructions of more complex sets, the abstract world of sets organizes itself into an *iterative hierarchy.* The iterative hierarchy is a connect-the-dots network (an abstract structure) in which the dots are sets and the connections are instances of the membership relation. Since many excellent books describe the iterative hierarchy, it will not be discussed in detail here. The iterative hierarchy is defined by three informal rules. The *initial rule* says there is an initial generation of sets. This initial generation is the empty set. The *successor rule* says that every generation of sets is surpassed by a greater successor generation of sets. Every successor generation is the *power set* of its previous generation. The power set of S is the set of all subsets of S. For instance, the power set of {A, B} is {{}, {A}, {B}, {A, B}}. The *limit rule* says that every infinite progression of ever greater generations of sets is surpassed by a greater limit system of sets. The limit system is the union of all the generations of sets in the progression of which it is the limit. The iterative hierarchy of sets is utterly topless.

Mathematicians have shown how to use sets to construct other types of mathematical objects, such as numbers and functions. One way to use sets to define the natural numbers states that zero is the empty set and every number *n* is the set of all numbers less than *n*. Thus one is the set containing exactly zero; two is the set containing exactly zero and one; and so it goes. Formally, 0 is {}; 1 is {0}; 2 is {0, 1}; and so it goes. Digitalists use set-theoretic constructions to define programs (Sections 20 and 33). Sets are the building blocks for the programs run by gods, cells, and bodies. Souls are made of sets. Digitalists also use set-theoretic constructions to define abstract computations. Sets are the building blocks for abstract models of program-execution.[1] For example, there are abstract cellular automata, built entirely out of sets. These abstract cellular automata are abstract universe-forms. These abstract universe-forms motivate *Pythagoreanism.*

The ancient Pythagoreans argued that mathematical objects are the only really existing things. And various modern philosophers and scientists have followed them. Among philosophers, Quine famously argued that all existing objects are ultimately made of sets (1976, 1978). Among scientists, the modern Pythagoreans include Schmidhuber (1997), Tegmark (1998, 2008), and Moravec (2000: ch. 7). All these Pythagoreans know that abstract universe-forms can be built entirely out of sets. These abstract universe-forms are purely mathematical models of physical theories. But these Pythagoreans argue further that if some abstract universe-form contains *self-aware substructures* (SASs), then those SASs perceive themselves and their surroundings as concrete. The Pythagoreans say that we are purely set-theoretic structures, inhabiting an abstract universe-form.

But most philosophers and scientists have not gone down this Pythagorean road. They have not been content to say that reality consists of the abstract background only. Against Pythagoreanism, many philosophers and scientists affirm *Platonism.* Platonists argue for the existence of both abstract mathematical objects and concrete physical things. Digitalists are Platonists. But Platonism (like any philosophy which involves concrete things) raises a terrifying *Question.* On the one hand, the abstract background of mathematical objects exists with an absolute necessity. It makes no sense to ask why it exists – it needs no explanation. On the other hand, concrete things do require explanations. And even if some things exist with concrete necessity (like gods), it is always reasonable to wonder *why* they exist. This is the Question: why are there any concrete things rather than none? Why is there something rather than nothing? The answer to the Question obviously cannot involve any concrete thing. It cannot involve any god or physical thing.

127. The eruption

Any answer to the Question of concrete existence must lie entirely within the abstract background. It can involve only abstract objects. Sets are abstract objects, but they cannot help here – they are utterly inert, eternally frozen in their austere majesty. Fortunately, the abstract background contains other objects: it contains *propositions*. Propositions are purely logical meanings. Our thoughts and sentences represent those meanings. The laws of nature are propositions. Since various evidence-based arguments justify the existence of propositions, digitalists affirm that they are natural objects.

The abstract background of propositions contains many *axiological requirements*, which are demands based on values. Among persons, axiological requirements manifest themselves as moral demands, duties, or obligations. Rules like "Do unto others as you would have them do unto you" and "You ought to keep promises" express moral demands. But moral demands are not the only axiological requirements. *Axiarchists* say the abstract background contains ontological demands, demands which it satisfies by bringing concrete things into existence. They say that value is creatively effective.[2] Early axiarchists include Leslie (1970, 1979, 1989, 2001) and Rescher (1984, 2000).

A digital version of axiarchism is briefly sketched here. This digital axiarchism is clearly inspired by the axiarchisms of Leslie and Rescher; however, this is not the place to either compare or contrast digital axiarchism with their older axiarchisms. Many analytic philosophers will find this brief sketch of digital axiarchism far too fuzzy or metaphysical to take seriously. Nevertheless, even those extremely careful analysts must answer the Question. Digitalists are free to explore their own axiarchic answer. And, if that answer is too fuzzy, then future digitalist philosophy must make it more precise. For digitalists, the creative power of value is expressed by the *Axiarchic Principle*, which states that for any proposition p, if it is axiologically required that p, then p.[3]

Of course, merely stating the Axiarchic Principle does not make it true. An argument is needed to justify the truth of the Axiarchic Principle. The *Axiological Argument* involves two key premises. The first states that the Axiarchic Principle is the best of all possible principles. And this is an analytic truth. What principle could be better than the one that asserts the satisfaction of every axiological demand? A modern Platonist, who wishes to avoid the woolly mysticism of the old Neoplatonists, can say that the Axiarchic Principle is the *Form of the Good* – the Axiarchic Principle is the *good itself*.[4] So the first premise is not problematic. The second premise states that any true principle is better than any false principle. And, once again, this premise is analytically true: it is analytically true that truths are better (intrinsically better) than falsehoods.

The *Axiological Argument* now runs like this: (1) The Axiarchic Principle is the best of all possible principles. (2) Any true principle is better than any false principle. (3) There are many true principles (like one plus one is two, hydrogen has one proton, and so on). (4) Suppose that the Axiarchic Principle is false. (5) If the Axiarchic Principle is false, then some principles are better than it. (6) But then it is not the best of all possible principles. (7) And since supposing the falsity of the Axiarchic Principle leads to a contradiction, the Axiarchic Principle must be true. This line of reasoning follows the pattern of Anselm's Ontological Argument. And, just as many philosophers regard that Ontological Argument as mere word play, so many philosophers will regard this Axiological Argument with equal skepticism. Nevertheless, it is an argument. Accepting this argument, digitalists conclude that the Axiarchic Principle is true. This truth is the *eruption* of abstractness into concreteness; it is the unfolding of all the abstract meaning that is enfolded in the Axiarchic Principle.[5] But the Axiarchic Principle is goodness itself.

Of course, it remains to define the axiologically required propositions. For digitalists, the ultimate axiological requirements are those which *ontologically* ought to be true, which express the duties of being towards itself. They hold at the deepest level of existence; but any propositions which hold at the deepest level of existence are the axioms which define the most inclusive system of existing things; hence the ultimate axiological requirements define the contents of the maximally-inclusive system of things; but the maximally-inclusive system of things is nature. The ultimate axiological requirements fill nature with value by axiomatically asserting the existence of valuable things.

Obviously, it is ultimately axiologically required that some valuable things exist; but the Axiarchic Principle states that all axiological requirements are true; hence some valuable things do exist. Of course, some abstract objects have values (these valuable abstract objects include true propositions and beautiful mathematical forms). For Platonists, the concrete instantiation of abstract forms increases the value of nature. Since the Axiarchic Principle is universal, it entails that those valuable concrete instantiations also exist. Hence valuable concrete things exist. Why is there something rather than nothing? Because the Axiarchic Principle entails that all valuable concrete things exist.[6] But what are those valuable concrete things? For digitalists, all concrete things are computations, hence all valuable concrete things are computations. The value of any computation is its intrinsic value, and its intrinsic value is its logical density (see Section 73).[7]

For digitalists, value is cumulative. Consequently, three rules define the class of valuable things. These three rules are the ultimate axiological requirements, and all other axiologically required propositions are derived from them. The *initial rule* states that the least valuable computation ought to exist. The *successor rule* states that if any valuable computation exists, then every improvement of that computation ought to exist. It states that for every computation, for every way to improve that computation, it is axiologically required that there exists some successor computation which is improved in that way. The *limit rule* states that if any progression of ever better computations exists, then every improvement of it ought to exist. It states that for every progression of ever better computations, for every way to improve that progression, it is axiologically required that there exists some limit computation which is improved in that way.

The Axiarchic Principle entails that these rules are true. And the richness of the abstract background (which is constrained only by consistency) ensures that there are always ways to improve every computation and progression of computations. Hence the totality of these ever more valuable computations exists. Since all these computations are concrete things, it follows that the concrete foreground, existing over the abstract background, is a *plenum* of value.[8] It contains all minimally valuable things; it is closed under the improvement relation; it is therefore maximally value-inclusive. Of course, every computation in that plenum is surpassable. But surpassability is defectiveness; hence while every computation in the plenum is defective, the plenum alone is unsurpassably good.[9] Within the plenum, everything that *ought* to be true *will be* true.[10]

128. Concreteness

Since the rules that define the plenum apply to all computations, they apply to the most basic computations. They apply to the computations that do not supervene on any deeper computations. They apply to the hardware on which all other computations run as software processes. But these hardware computations are just the digital gods. Hence the axiarchic rules entail the *epic of theology* (Section 77).[11] And since the rules in that epic are true, the divine hierarchy exists – the Great Tree of gods exists. The Great Tree of gods is the pleroma; but the plenum of computations exceeds the pleroma.

According to the epic of theology, there exists exactly one minimally valuable god, namely, Alpha. It supports the minimally valuable universe, the empty universe. As gods grow in value, they gain logical density. But as they gain density, they become internally more complex.[12] Any such increase in complexity eventually entails the emergence of increasingly complex internal computations, software processes running on the gods. It entails that the universes running on these gods become more complex themselves. They contain increasingly complex physical structures. Hence the axiarchic rules entail the rules for the improvements of universes. They entail the *epic of cosmology* (Section 83). And, since the rules in that epic are true, the cosmological hierarchy exists.

As the universes grow more valuable, they also gain in logical density. But as they gain density, they too become internally more complex. Any such increase in complexity eventually entails the emergence of increasingly complex internal computations, software processes running inside of these universes. These software processes are physical things, such as points, protons, plasmas, planets, paramecia, plants, puppies, primates, and people. Hence the axiarchic rules entail the rules for the *epic of physics* (Section 89). And since lives are processes, they entail the *epic of biology* (Section 95). They entail that every life sits at the root of its own endlessly ramified tree of life.

Every life is saved within its tree of life; its tree of life is its salvation. The salvation of any life is a structure which contains that life and which is closed under the improvement relation. It is defined by the expected three rules (initial, successor, and limit). The rules for any tree of life, including your tree of life, follow from the axiarchic rules. They follow from the Axiarchic Principle, which is the abstract root of all concreteness; they follow from the eruption of abstractness into concreteness; they follow from the concrete unfolding of all the abstract meaning that is enfolded in goodness itself.

The Axiarchic Principle, which expresses the computational *unity* of nature, also expresses its computational *self-transcendence*. According to this self-transcendence, every consistently definable concrete structure is surpassed by every consistently definable superior version of itself. For digitalists, this self-transcendence is the best explanation for the existence of any concrete structures at all. The application of this self-transcendence to *living structures* entails universal salvation. It entails life after death.

Notes

I Ghosts

1. Famous mummies include those of the Egyptian pharaohs; those of people frozen in ice (such as Otzi the Iceman); and those of people buried in bogs (such as Lindow Man). The DNA in the mummy of Otzi has been sequenced (Keller et al., 2012).
2. One ancestor of digital ghosts is the Memex system (Bush, 1945). Another is the Teddy (Norman, 1992: ch. 6. The *recorded personalities* or *constructs* in William Gibson's *Neuromancer* are also like digital ghosts (1984: 76–7). Ghosts play central roles in Gibson's *Mona Lisa Overdrive* (1988). Nozick described intelligent programs that know your life history and that can simulate your personality (1989: 24–5).
3. Technologies for digital ghosts fall under the *Capture, Archival and Retrieval of Personal Experiences* (CARPE). One CARPE project is MyLifeBits (see Gemmell et al., 2002, 2003). For more on CARPE, see Steinhart (2007a).
4. As of June 2013, when Facebook is informed of your death, it changes the status of your account to *memorial*. Memorial timelines preserve most of their original content, but they are accessible only to friends and do not appear in searches.
5. Members of the *Quantified Self* movement digitally log many aspects of their daily lives, especially their physiological data (Swan, 2012).
6. Your second-generation ghost can include models of your personality based on your answers to long psychological exams (Bainbridge, 2003). It can include the ways you have tuned recommendation engines (Shafer et al., 2001). Those engines learn and store your preferences in books, music, art, movies, clothes, food, dating, and mating.
7. The classical digitalists argue that mentality is entirely computable (see Moravec, 1988, 2000 in their entirety). More specifically, see Moravec (2000: 72–88, 121–4) and Kurzweil (1999: ch. 3; replies in 2002; 2005: chs. 3 & 4).
8. An accurate digital ghost can pass a *personalized Turing test*. When your friends and family interact with your ghost (via some electronic interface), it can convince others into thinking that it is you. See Kurzweil (2005: 383).

II Persistence

1. The concept of pipes as causal arrows is inspired by Salmon's (1984) mark-transmission theory of causality and Dowe's (1992) conserved quantity theory. Pipes transmit bits of information and they conserve program functionality.
2. The program shown in Figure 3 is Wolfram's Rule 150 (it is 10010110).

3. Pipes are programs that transmit existence (they are if-then rules). Any pipe has this form: if *there exists* some previous object with its previous properties, then *there exists* some next object with its next properties. The programming of the pipe defines the relation between the previous properties and the next properties.

4. If some previous stage is linked by an updating pipe into its next stage, then the next stage is a successor of that previous stage. A stage *y* is a successor of stage *x* if and only if *p* is the program (I, S, O, F, G); *x* has the configuration (p, i, s, o); *y* has the configuration (p^*, i^*, s^*, o^*); p^* is identical with p; the input i^* comes from the environment; the state s^* is $F(i, s)$; and the output o^* is $G(i, s)$. All the stages in some process are linked by updating pipes if and only if each next stage is a successor of its previous stage.

5. The classical digitalists often write about the game of life. The classical digitalists include writers like Edward Fredkin, Hans Moravec, Ray Kurzweil, and perhaps Frank Tipler. See Moravec (1988: 151–8); Tipler (1995: 37–8); Kurzweil (2005: 520).

6. Digitalists affirm the *Indiscernibility of Identicals*, which states that for any things *x* and *y*, if *x* is identical with *y*, then for every property *p*, *x* has *p* if and only if *y* has *p*. Having parts is like having properties. Hence this principle also implies that for any things *x* and *y*, if *x* is identical with *y*, then for every thing *z*, *z* is a part of *x* if and only if *z* is a part of *y*. The Indiscernibility of Identicals is logically necessary.

7. Those who affirm identity through time are *endurantists*. They say that things endure. But many philosophers have denied that there is any identity through time. They have attacked endurantism. Heraclitus was probably the first philosopher to attack endurantism: "You cannot step into the same river twice." Since Heraclitus, many other philosophers have taken up the attack (see Quine, 1950; Browning, 1988; Heller, 1990: ch. 1). The arguments against endurantism are brilliantly developed by Sider (2001) and Hawley (2001). Like Heraclitus, digitalists argue against endurantism.

8. For a good collection of readings on persistence, see Haslanger & Kurtz (2006). For discussions of persistence in the game of life, see Steinhart (2012a).

9. For arguments that our universe is a cellular automaton, see Fredkin, Landauer, & Toffoli (1982); Fredkin (1991); Zeilinger (1999); Fredkin (2003). Still, it does not seem likely that our universe is a cellular automaton. For other ways to think of our universe as a network of computers, see Wolfram (2002); Kurzweil (2005: 85–94; 518–22).

10. Aristotelian physics involves only 3D substances interacting in a single universal present. For conflicts between special relativity and Aristotelian endurantism, see Balashov (2000a, 2000b), Hales & Johnson (2003), Sider (2001: ch. 4.4).

11. A discrete process is a function *f* from some ordinal K to some set of stages such that (1) $f(m)$ is earlier than $f(n)$ if and only if *m* is less than *n*; and (2) for every successor ordinal $n+1$ in K, the stage $f(n)$ persists into its successor stage $f(n+1)$; and (3) for every limit ordinal *h* in K, the series $\langle f(i) | i < h \rangle$ persists into its limit stage $f(h)$.

12. On the one hand, the classical digitalists have many discussions of persistence (Moravec, 1988: 116–20; Tipler, 1995: 227–40; Kurzweil, 2005: 383–6). On the other hand, exdurantism is not developed until Sider (1996, 2001)

and Hawley (2001). So Moravec and Tipler cannot have heard about it. And it does not seem fair to expect Kurzweil to have studied Sider and Hawley. Still, exdurantism is the theory of persistence most compatible with Kurzweil (2005: 383–6).

13. Kurzweil identifies himself with his body-pattern (2005: 385–6). He argues that the constant replacement of his parts "means the end of me even if my pattern is preserved" (385). This leads him to wonder "So am I constantly being replaced by someone else who just seems a lot like the me of a few moments earlier?" (385). On the basis of his own reasoning, his answer must be *yes*. Sadly, he then indicates that he will ignore his reasoning (386). He should have stuck with his reasoning – it was correct.

14. One pipeline may *fission* into many. Fission typically preserves programs. If the process Original fissions into the processes Left and Right, then one program runs through Original plus Left and also through Original plus Right. Hence Original plus Left and Original plus Right are two overlapping processes. Many pipelines may fuse into one pipeline. Fusion typically fails to preserve program functionality – the fused programs blend into one new program. If Right and Left fuse into Later, then Right and Left each end with the fusion, and Later is an entirely new process.

15. Fission (that is, twinning) negates identity through time. No thing can retain its identity through fission. Suppose A divides into B and C; the symmetry of the division and the preservation of form entail that A persists into B and A persists into C; but if this persistence is identity, then A is B and A is C; and, since identity is transitive, B is C; but that is wrong: B is not C. Hence fission shows that there is no identity through time. But fission does not negate persistence or continuity through time.

16. Cells divide naturally. But philosophers and digitalists have discussed artificial body-fission (Parfit, 1985: 199–201; Moravec, 1988: 116–20; Kurzweil, 2005: 383–5). The best way to make sense out of body-fission is via exdurantism.

17. Someone may wish to argue that an organism must be biologically autonomous; yet no embryo has such autonomy; hence no embryo is an organism; on the contrary, it is merely an organ in the maternal body. On that view, early embryonic stages lack the values and rights of organisms. Nevertheless, on that view, later embryonic stages gain the values and rights of organisms as they gain biological autonomy.

18. The classical digitalists use the term *pattern* to refer to the program being run by a body (Moravec, 1988: 116–20; Tipler, 1995: 227–40; Kurzweil, 2005: 383–6). At times they seem to say that persons or bodies are identical with their patterns. But that is absurd: bodies and persons are concrete particular things; they are not abstract patterns.

19. Counterpart theory has its roots in the logic of possibility. It has its roots in the theory that there are other possible universes populated by other possible people – other possible versions of yourself. The logic of possibility is *modal logic*. Counterpart theory is used in modal logic to solve problems involving *trans-world identity* (Lewis, 1968, 1986). Counterpart theory is used in temporal logic to solve problems involving *trans-time identity* (Lewis, 1973: sec. 5.2; Sider, 1996, 2001). Digitalism uses counterpart theory to solve problems involving *trans-death identity* (see Steinhart, 2008).

20. Parfit says that the poem comes from the *Visuddimagga*. He takes it from R. Collins, *Selfless Persons* (New York: Cambridge University Press, 1982), p. 133.

III Anatomy

1. All earthly cells are only finitely complex. Cells with infinite complexity exist in future limit revisions of the earthly universe (see Chapters VIII and IX).
2. Of course, since the theory of differential equations is highly advanced, it may be very *convenient* to use differential equations to describe biological systems. But those differential equations are merely the idealizations of difference equations.
3. For descriptions of cells as networks of nano-machines, see Barabasi & Oltvai (2004); Szallasi et al. (2006); Alon (2007); Helms (2008).
4. There are many ways to provide abstract nanite networks with concrete biological interpretations (Bornholdt, 2005; Fisher & Henzinger, 2007). On some interpretations, each nanite corresponds to an individual molecule or organelle (to one protein, gene, unit of RNA, proteasome, or ribosome). On others, each nanite corresponds to the concentration of a certain species of molecule. Some interpretations regard the relations among nanites as Boolean operations (Bornholdt, 2008; Helikar et al., 2008; Morris et al., 2010) while others regard them as differential equations (Conrad & Tyson, 2006). Of course, these differential equations are idealized difference equations.
5. Motifs and modules are examples of higher-level nanites realized by networks of lower-level nanites (Hartwell et al., 1999; Tyson et al., 2003).
6. For cellular pattern recognition and decision making, see Bray (1990, 1995); Bhalla & Iyengar (1999); Scott & Pawson (2000); Hellingwerf (2005); Helikar et al. (2008).
7. A cell is nondeterministic if and only if its associations are relations but not functions (they are one-to-many or many-to-many). Digitalists incorporate nondeterministic things by allowing time to branch. When time branches, things fission. Branching time entails a 4D theory of persistence. However, since universes with branching times merely add technical complications, they are not considered here.
8. Formally, cell y is a successor of cell x if and only if p is the program (I, S, O, F, G); x has the configuration (p, i, s, o); y has the configuration (p^*, i^*, s^*, o^*); p^* is identical with p; the input i^* is in I; the state s^* is $F(i, s)$; and the output o^* is $G(i, s)$.
9. All cells considered here are *eukaryotic*. A eukaryotic cell has a nucleus that houses its genetic material. The cells of plants and animals (including humans) are eukaryotic. Bacterial cells are not eukaryotic but *prokaryotic*.
10. Bions include E-CELL (Tomita et al., 1999; Tomita, 2001) and Virtual Cell (Lowe & Schaff, 2001; Moraru, 2008). Many striking advances have recently been made in whole cell simulation (Roberts et al., 2011; Karr et al., 2012).
11. Some writers have argued that the brain is a quantum computer whose powers therefore exceed those of classical computers (universal Turing machines). Against this, it has been argued that quantum computation is irrelevant to cognition (Litt et al., 2006). For a deflationary view of quantum computing, see Aaronson (2008).

12. It is easy to define *reference* for immune memories: the DNA in the VDJ genes in B-cells is a string of symbols that determines the protein structure of an antibody; the antibody is effective against its antigen because they fit like a lock and key. Immune memory models correspondence and causal theories of reference.

13. Sagan (1995: 987) says there are about 10^{14} cells in your body. Each cell contains about 10^{12} bits of information. The number of cells in your body times the number of bits per cell yields 10^{26} bits in your body. Quantum mechanics entails an upper bound (the Bekenstein bound) on the amount of information that can be stored in any amount of mass (Bekenstein & Shiffer, 1990). Moravec (2000: 166) uses the quantum mechanical theory of information to calculate the maximum amount of information in any human body. He says we each contain about 10^{45} bits—a small finite number.

14. Burks (1973) gives a nice argument that every body is a finitely complex machine. Tipler writes that a human body can be "in one of $10^{\wedge}(10^{\wedge}45)$ states at most, and can undergo at most $4^{*}10^{\wedge}53$ changes of state per second ... a human being is a finite state machine and *nothing but* a finite state machine" (1995: 31).

15. As was the case for cells, a body is nondeterministic if and only if its associations are relations but not functions. Digitalists incorporate nondeterministic bodies by allowing time to branch. When time branches, nondeterministic bodies fission. Branching time entails a 4D theory of persistence. However, since universes with branching times merely add technical complications, they are not considered here.

16. Specialized hardware for cell-network simulation includes designs like field programmable gate arrays and neuromorphic chips (see Boahen, 2005).

17. The field of massively parallel supercomputers evolves rapidly. For a recent overview, see Narayan (2009). Much of what is said here will surely soon be obsolete.

18. The standard definition of persons lists various psychological and moral abilities (intelligence, rationality, linguistic competence, self-awareness, second-order beliefs and desires, moral agency, and so on). See Locke (1690: II.27.9); Frankfurt (1971); Olson (1997: 4–5, 102–4); Hudson (2001: ch. 4, sec. 3).

19. Human persons include at least all members of the species *Homo sapiens*. Whether they are human or not, it is likely that Neanderthals are persons. Perhaps chimpanzees, gorillas, early hominids, dolphins, elephants, corvids, and parrots are persons. Possible non-human persons include rational moral robots and rational moral extraterrestrials. For theists, the Personal Omni-God (POG) is a nonhuman person.

20. It has been argued that you are *constituted* by your body but are not *identical* with it (Baker, 2000). Hudson (2001) successfully refutes such constitutionalism. But even if it were true, it would make no difference to our arguments.

21. Digitalism says that for every x, if x is a person, then x is a body. It is not the case that for every x, if x is a body, then x is a person. Normal healthy adult bodies are persons. Probably bodies gradually gain personhood as they move towards adulthood. But none of the present arguments depend on which bodies are persons.

22. The classical digitalists all reject Socratic-Cartesian mind–body dualism. Moravec says "mind is entirely the consequence of interacting matter" (1988: 119). And he explicitly rejects Cartesian mind–body dualism (2000: 121–4). Tipler rejects dualism: "a human being is a purely physical object, a biochemical machine completely and exhaustively described by the known laws of physics" (1995: 1). Kurzweil rejects dualism (1999: 55–65; Kurzweil, 2002: 191–4). He says that consciousness "does not require a world outside the physical world we experience" (2002: 214). The classical digitalists sometimes endorse something like a software–hardware dualism or form–matter dualism. However, those dualisms are not Socratic-Cartesian mind–body dualisms.

23. Your digestive tract contains your undigested food and all your gut bacteria. About nine out of ten cells in your body are gut bacteria. Kurzweil (2005: 386) says that the material in your gut is not part of your body. But why not? Your gut bacteria are essential to your life. Your gut bacteria play important roles in cognition – they control gene expression in the brain and secrete many chemicals that influence the nervous system. The lesson is that *every human body is an ecosystem.* The core of this ecosystem consists of cells with your DNA; but the core is merely a small proper part of the ecosystem.

24. Several factors imply an entirely physiological basis for NDEs: (*a*) people who have had NDEs typically have significant sleep disorders involving the intrusion of dream states into awareness (Nelson et al., 2006); (*b*) people who have had NDEs typically have significantly different kinds of temporal lobe functionality (Britton & Bootzin, 2004); (*c*) the physiology of dying appears to trigger large releases of glutamate in the brain, which affects the N-methyl-D-aspartate receptor (Bonta, 2003); (*d*) people who take ketamine have psychedelic experiences much like NDEs (Jansen, 1997).

25. OBEs can be reliably produced by causing discrepancies in visual and tactile body-perception. OBEs are entirely natural illusions in which the brain is deceived by incongruous stimuli (Ehrsson, 2007; Lenggenhager, 2007). OBEs are naturally produced by migraines and by epileptic seizures. They can be produced by electrically stimulating certain parts of the brain (De Ridder et al., 2007).

26. Some classical digitalists thought that if bodies do not remain the same through time, then persons cannot be bodies. However, when he argues against "body-identity", Moravec (1988: 116–22) confuses persistence with identity. He also confuses indiscernibility with identity. Astonishingly, he fails to recognize that identity is transitive. His discussion of identity is incoherent. Similar mistakes are made by Tipler (1995: 227–40). These writers are not rejecting the identity of the person with the body; they are rejecting *body-endurantism.* All digitalists reject all types of endurantism.

27. The soul is the form of the body. But the soul is not a *substantial form* (contra Aquinas). The soul is merely the abstract functional blueprint of the body. Just as the blueprint of an airplane can exist without the airplane of which it is the blueprint, so the soul can exist without the body. But the blueprint of an airplane is not an airplane. It cannot fly. Apart from the body, the soul is equally inert. It can neither live nor think.

28. Many writers say that the soul is to the body as software is to hardware (Reichenbach, 1978; Hick, 1976: 281–3; Polkinghorne, 1985; Barrow &

Tipler, 1986: 659; Mackay, 1997). Tipler writes that "the human 'soul' is nothing but a specific program being run on a computing machine called the brain"(1995: 1–2).

29. A linguistic analogy can be used to talk about genetic correction. Any genotype is like a sentence in a language. A language has grammatical rules. The parts of a sentence that are grammatical entail corrections to the parts of the sentence that fail to be grammatical. Hence the parts of any genotype that are grammatical entail corrections to the parts that are not grammatical. Every genotype entails its corrected versions.

30. Platonism (as used here) is realism about abstract objects. More precisely, it is realism about mathematical objects (such as sets, numbers, and functions). Moravec endorses Platonism (1988: 178; 2000: 196–8) and Tipler endorses it (1995: 213).

31. On this definition, empirical justification goes far beyond empirical testability. Many statements are empirically justified but are neither empirically verifiable nor empirically falsifiable. Hence the naturalism developed here is not positivism.

32. Principles of reason include the Identity of Indiscernibles; the thesis that there are no infinitely descending dependency chains; the Principle of Sufficient Reason (PSR); the Principle of Plenitude (PP); and Ockham's Razor. The Identity of Indiscernibles shows up in the extensionality axiom of standard Zermelo–Fraenkel–Choice (ZFC) set theory; the rejection of infinitely descending dependency chains shows up in the ZFC axiom of foundation. Kane (1986) argues that the PP and PSR are used in science.

33. Nature is an absolutely rich relational structure. It is that relational structure than which no greater is consistently definable (logically possible). It is a connect-the-dots network that is so big that embedding it in any bigger network is contradictory.

34. It excludes phlogiston, *élan vital*, and luminiferous ether. It excludes enduring substances, immaterial minds, and inconsistent objects like four-sided triangles and the set of all self-excluding sets. It excludes mythological objects like imps and the old pagan gods. It excludes occult energies, vibrations, spirits, and so on.

IV Uploading

1. Digitalists are interested in uploading persons. On the erroneous assumption that persons are brains, some classical digitalists have focused on brain scanners (Moravec, 1988: 109–10; Kurzweil, 2005: 198–201). However, since persons are bodies rather than brains, full body scanners are required to upload persons.

2. For scanners in teleportation systems, see Wiener (1954: 95–102); Parfit (1985: 199–201); Moravec (1988: 117–18). *Star Trek* famously illustrates teleporters.

3. Although the body-file produced by body scanning is a kind of digital ghost, it is distinct from the digital ghosts described in Chapter I. It contains only your natural memories. If your artificial digital ghost were used to augment the memory of your avatar, the result would not be mere uploading. It would be a kind of enhancement.

4. Some effort has gone into making biological simulators that model organs like the heart (Noble, 2002) and the liver (Yan et al., 2008). But most effort has focused on building animats that approximate the brain (Markram, 2006; Djurfeldt et al., 2008; Izhikevich & Edelman, 2008; Sandberg & Bostrom, 2008; King et al., 2009). And the Physiome Project aims at whole-body simulation (Hunter & Borg, 2003).

5. For information on Second Life, see Rymaszewski et al. (2007) and Malaby (2009); for World of Warcraft, see Bainbridge (2010).

6. Moravec describes a computer able to simulate the entire surface of the earth at the atomic level of detail (1988: 122–4). Tipler describes an infinite computer able to simulate any finite physical system (1995: 248–9, 265, 462, 505). Sandberg (1999) describes three technically possible super-computers that could be used as terraria.

7. The Millennium Simulation simulates the outer sphere of stars and other astronomical structures (Springel, 2005). The degree of realism in the Millennium Simulation is probably sufficient for the outermost shell in a terrarium. Google Earth is a simulation of the surface of the earth. Circa 2013, it consists of 3D data for projection onto 2D screens with a resolution between 15 and 2.5 meters per pixel.

8. Many writers have discussed uploading (see Dennett, 1978: ch. 17; Brown, 1990; Kurzweil, 1999: ch. 6; Broderick, 2001: ch. 6; Chalmers, 2010: 41–63). Moravec (1988: 110–12) describes the transference of your brain-structure into a robotic body. Such transference (he calls it "transmigration") is distinct from uploading.

9. Although Moravec is a materialist about the mind, he often uses dualistic language when he writes about uploading (1988: 108–112; 2000: chs. 5 & 6). And Goertzel & Bugaj (2006) use dualistic language when they write about uploading. However, since dualism is false, such language is misleading. It is misleading to talk about *mind uploading* or about *advanced substrate independent minds*. It is better to talk about *brain uploading* (Kurzweil, 2005: 198–204); or about *whole brain emulation*. Of course, since persons are bodies, it is best to talk about *body uploading* or *whole body emulation*.

10. Digitalists affirm that artificial computing machines can have minds that are functionally equivalent to natural human minds. Artificial machines can be aware and self-aware. See Moravec (2000: 72–88); Kurzweil (1999: 55–63, 116–18). See Kurzweil's replies to his critics in Kurzweil (2002: chs. 6–10). And see Kurzweil (2005: 376–82). Within Kurzweil (2005: ch. 9, see especially 435–55, 458–68, 473–83).

11. Persons are bodies; hence persons go where their bodies go (Steinhart, 2001). Persons do not go with brains (contra Puccetti, 1969). To upload a person, it is therefore not sufficient to merely upload their brain. Uploading the brain is equivalent to producing a brain in a vat. A brain in a vat lives in a hallucinated reality (much like a Cartesian mind alone with its Evil Demon). Price (1956) says that the afterlife may be just an enormous hallucination of a disembodied mind. Such a mind is not a person. Merely uploading the brain does not transport a person into cyberspace.

12. Bodies in an uploading span are processmates if and only if bodies persist through teleportation. Following Parfit (1985), many philosophers argue that bodies do persist through teleportation. Anybody who says personal

persistence is psychological continuity should agree that all bodies in a span are processmates.

13. Another type of body uploading does not involve destructive scanning – it's like an MRI scan, which leaves your old organic body just the way it was. However, even in this case, there is no biological continuity from your old body into your new body.

14. For resurrection as replication, see Parfit (1971); Hick (1976: ch. 15); Polkinghorne (1985: 180–1, 2002); Reichenbach (1978: 27); Mackay (1997: 248–9).

15. The digitalist definition of resurrection says that for any body x, x will be *resurrected* if and only if x is an organic body and x will be a resurrection body. Something like this can be found in the New Testament. Paul says: "Just as we have borne the image of the man of dust, we shall also bear the image of the man of heaven" (1 Cor 15: 49; RSV). The organic body is the image of the man of dust (Adam) while the resurrection body is the image of the man of heaven (Christ). The digitalist definition is more general.

16. One well-known objection to counterpart theory says that we don't care about our counterparts (Kripke, 1980: 45). But Lewis (1986: ch. 4) and Miller (1992) give the correct reply: we ought to care about our counterparts because many of our present properties depend on them. Depending on whether or not you have a future counterpart, right now either you *will live a better life in a better future* or else you *will not*. And you ought to care about which of those properties you presently have.

17. For universe-design in science fiction, see Dozois (1991), Ochoa & Osier (1993), and Gillett & Bova (2001). For universe-design in games, see Salen & Zimmerman (2003) and Schell (2008). Universe-design in games is most relevant here.

18. Tipler says that we could be resurrected in a world that is "as close as logically possible to the ideal *fantasy* world of the resurrected dead person" (1995: 241). But human fantasy is so incoherent that it is difficult to make sense of this claim. Our fantasies are often realized in our dreams; but in our dreams we are alone; fantasy is solipsism.

19. Moravec (1988: 112–16) describes many transformations of an uploaded person that do not seem to be consistent with any type of human continuity.

20. Perhaps this recalls Richard Brautigan's techno-arcadian poem "All Watched Over by Machines of Loving Grace".

V Promotion

1. Medieval writers distinguished between *accidentally* and *essentially* ordered causes (Wengert, 1971; Harrison, 1974). Accidentally ordered causes are efficient causes while essentially ordered causes are supportive or sustaining causes. Sequences of efficient causes may be infinitely regressive; however, sequences of supporting causes must be well-founded. The Second and Third Ways involve supporting causes rather than efficient causes (Davis, 1992). For digitalists, software objects are supported by deeper software objects; virtual computations are supported by deeper computations. But stacks of ever deeper computations are necessarily well-founded – after finitely many

steps, they bottom out in hardware. These stacks are held up by supportive causes.

2. Opponents of cosmological arguments typically insist that support chains can be infinitely descending – it's turtles all the way down. But Cameron (2008) argues that it is reasonable to grant axiomatic status to the thesis that there are no infinitely descending support chains. Here digitalists agree with Cameron. This is a *metaphysical axiom of foundation:* all ultimate support chains are well-founded. Of course, digitalists happily agree that many virtual support chains are infinitely descending. But those virtual dependency relations are defined over deeper systems whose support relations are well-founded. The infinitely regressive integers are defined over the well-founded natural numbers.

3. Models of computation that exceed classical Turing machines include Giunti machines (Giunti, 1997); accelerating Turing machines (Copeland, 1998a, 1998b); various machines operating on transfinite ordinals (see the articles by Hamkins and by Koepke); and various continuous computers (Moore, 1996; Blum et al., 1998).

4. Although the Simulation Argument is naturalistic, Bostrom points out that it has many potentially religious consequences. Suppose that our universe is running on an Engine built by some Engineers. Bostrom (2003: 253–4) writes that the Engineers "are like gods in relation to the people inhabiting the simulation: the [Engineers] created the world we see; they are of superior intelligence; they are 'omnipotent' in the sense that they can interfere in the workings of our world even in ways that violate its physical laws; and they are 'omniscient' in the sense that they can monitor everything that happens."

5. A large literature exists on the Fine Tuning Argument (FTA). Only a few landmarks can be mentioned here. For excellent introductions to the FTA, see Barrow & Tipler (1986) and Leslie (1989). For a presentation of the FTA that involves a universe-producing machine, see van Inwagen (1993: 130–1). There are difficulties with the FTA. Manson (2000) argues that there is no adequate definition of fine tuning. Colyvan et al. (2005) argue that the FTA cannot be supported by probability theory. But Koperski (2005) argues that there are definitions of fine tuning that satisfy those demands.

6. It is standard to say that something is God if and only if it is maximally perfect. Since the Engineers are not maximally perfect, they are not God. Since no individual Engineer is more perfect than the Engineers, no individual Engineer is God.

7. The lives of all humans (and all other rational creatures) are recorded by the Engine and stored in some great database. The database stores perfect physiological ghosts. It resembles the *Book of Life*, in which God has recorded all human lives (Exodus 32: 32; Psalms 56: 8; Psalms 69: 28; Daniel 12: 1; Luke 10: 20; Philippians 4: 3; Revelations 3: 5 and 20: 11–15). It also resembles the *akashic record* in occultism. And having a record in the database resembles the *objective immortality* of Hartshorne. However, unlike the Book of Life or the akashic record, the database is naturalistic.

8. For resurrection as *re-creation*, see Shorter (1962: 81–4); Sutherland (1964: 386); Hick (1976: 465); Forrest (1995: 58); Kundera (1999: part 5, sec. 16).

9. Paul says the resurrection body will be glorified (1 Corinthians 15). Augustine devoted large sections of *The City of God* to the resurrection body. See bk. 11 ch. 23; bk. 13 chs. 13, 16–20, 22, 23; bk. 14 chs. 3, 15–16, 19–26; bk. 16 ch. 8; bk. XX; bk. 21 chs. 2–3; bk. 22 chs. 10–21, 29, 30. Aquinas discusses the resurrection body in the Supplement to the *Summa Theologica* (Q. 80–5). These ancient and Medieval discussions of the resurrection body make very little scientific sense.

10. The first reference to nested simulations appears to be that of Wright: "any universal computer can simulate another universal computer, and the simulated computer can, because it is universal, do the same thing. So it is possible to conceive of a theoretically endless series of computers contained, like Russian dolls, in larger versions of themselves" (1988: 42). This reference occurs in an article about Edward Fredkin.

11. For convenience, it is said that our universe is on level zero. But if we run genuine universe simulations, or if aliens on other planets in our universe do, then our universe just moves up in the hierarchy; other numbers get shifted as needed.

12. Every finite engine both ultimately and directly depends on the Deity. Hence there are no infinitely descending dependency chains. The structure of the dependency relation is isomorphic to that of the greater than relation in the set $\{0, 1, 2, \ldots, \omega\}$.

13. The Deity is causally responsible for every universe at every moment of its existence. This is a doctrine of *continuous creation*. See Descartes, *Principles of Philosophy*, part 1, sec. 21; *Meditations*, 3, para. 31. For digitalists, continuous creation supports 4D theories of physical persistence (such as exdurantism, see Chapter II).

14. Any classical Turing machine has infinite memory. But it operates only at finite speeds and it can store only finitely many 1s in its memory. For truly infinitely powerful engines, consider *accelerating Turing machines* (ATMs). An ATM can store infinitely many 1s and it can perform infinitely many tasks in any finite time. ATMs can solve problems that cannot be solved by classical Turing machines (Copeland, 1998a, 1998b). They can perform supertasks. Hence the Deity is equivalent to an ATM.

15. For any finite n, if the density of universe U(n) is D, then the density of U($n+1$) is 2^D and if the speed of U(n) is S, then the speed of U($n+1$) is 2^S.

16. The hierarchy of engines looks like the old great chain of being (Lovejoy, 1936). It resembles the medieval hierarchy of celestial spheres (which were thought of as nested heavens). For example, in Dante's *Paradiso*, there are ten concentrically nested heavens. And the engineers are analogous to the medieval choirs of angels.

VI Digital Gods

1. Dawkins says that it is almost certain that our universe is not designed (2008: ch. 4). He allows that there is some small probability that it is designed (p. 136, contra p. 186). And he argues that *if* our universe is designed, *then:* (1) Our Designer is complex (178–9). (2) Since all complex things emerge at the ends of "graded ramps of slowly increasing complexity" (139), our

Designer "must be the end product of some kind of cumulative escalator" (186). (3) The series of ever more complex things on this escalator starts with some first cause (185). And (4) "[t]he first cause that we seek must have been the simple basis for a self-bootstrapping crane" (185).

2. The concept of the *world tree* is found in many religions and mythologies. For example, the world tree in Norse mythology is known as *Yggdrasil*.

3. Our Designer is a god; just as the ancestors and descendents of organisms are organisms, so the ancestors and descendents of gods are gods; since every object in the Great Tree is either a descendent of our Designer or a descendent of some ancestor of our Designer, it follows that every object in the Great Tree is a god.

4. Dennett's Principle of Accumulation of Design says "since each new designed thing that appears must have a large design investment in its etiology somewhere, the cheapest hypothesis will always be that the design is largely copied from earlier designs, which are copied from earlier designs, and so forth" (1995: 72).

5. Randomness within any universe entails that it includes nondeterminism. It entails that time branches in that universe and that things fission when time branches. Each path in the branching time is one member of the sample space over that universe. Any feature which varies across these paths is a random variable. The variants found on the paths define the probability distribution (or density function) of that variable.

6. Bower (1988) argues that the complexity of a genotype is proportional to the number of distinct cell types in its phenotype. Adami et al. (2000) define the complexity of an organism directly in terms of genetic information. Nehaniv & Rhodes (2000) provide an insightful mathematical analysis of biological complexity. More recently, it has been argued that the complexity of any genotype is the ratio of its non-protein-coding-DNA to its total amount of DNA (Taft, Pheasant, and Mattick, 2007).

7. It might be objected that infinite regressions of ever simpler things are possible; hence our escalator need not start with some initial simple thing. The reply is the *Argument against Infinite Regressions of Gods:* (1) Any objects which can participate in infinite regressions are higher-level objects which are defined in terms of more basic lower-level objects. For example, integers participate in infinite regressions, but integers are defined in terms of natural numbers; infinite sets permit infinite regressions of subsets, but those infinite sets are defined in terms of finite sets; and the membership graphs of non-well-founded set theories are modeled within well-founded ZFC. (2) Since gods are the lowest-level concrete things, they are not definable in terms of any lower-level concrete things. (3) Therefore gods cannot participate in any infinite regressions. Infinite gods may support internal infinite physical regressions; but gods are like the ordinal numbers. One might argue further that complexity itself is ultimately quantified by ordinal numbers.

8. For more on divine evolution, see Steinhart (2013). The Dawkinsian Algorithm is basically the Leibnizian Search for Hardware.

9. It may be argued that two simple gods are merely theologically simple, and that such simplicity may include enough complexity to distinguish them. But if there are two such theologically simple gods, then their common intersection is some simpler god.

10. The initial simple god (Alpha) is similar to the Dawkinsian first cause. Dawkins allows that there exists an ultimate first cause (2008: 184). He says that "The first cause that we seek must have been the simple basis for a self-bootstrapping crane which eventually raised the world as we know it into its present complex existence" (2008: 184–5). Dawkins denies that the first cause is God (1996: 77; 2008: 101, 184). And Dawkins states that, if our universe has a designer, then it is not the first cause (1996: 77).

11. Dawkins argues that the POG is complex (2008: 136, 151, 171–2, 176–180, 183–4). POG is a mind; all minds are complex (Hume, 1779: 70–1); hence POG is indeed complex. After all, POG is not the Plotinian One.

12. Complexity is not entropy. On the contrary, it is small for both regular and random objects. Thus measures of entropy like algorithmic incompressibility (*a la* Kolmogorov) and *thermodynamic depth* (Lloyd & Pagels, 1988) do not measure complexity.

13. Bennett says the logical depth (density) of a digital object (such as a string of 0s and 1s) is "the time required by a universal computer to compute the object in question from a program that could not itself have been computed from a more concise program" (1985: 216); or "the time required by a standard universal Turing machine to generate it from an input that is algorithmically random" (1988: 1); or the "execution time required to generate the object in question by a near-incompressible universal computer program" (1990: 142). These definitions are equivalent. Any definition of density for physical things requires some translation (a simulation relation) between bit strings and those things.

14. Bennett says that "a [dense] object is one that is implausible except as the result of a long computation" (1985: 223). And that "a logically [dense] or complex object would then be one whose most plausible origin, via an effective process, entails a lengthy computation" (1988: 3). And also that "a [dense] object … contains internal evidence that a lengthy computation has already been done" (1988: 5).

15. Elliot says "a thing has intrinsic value if it is approved of by a valuer in virtue of its properties" (1992: 140). Davison indicates that "Something is intrinsically valuable only if it would be valued for its own sake by all fully informed, properly functioning persons" (2010: book cover text). These are *not* definitions of *intrinsic* value. They are definitions of values conferred *extrinsically* on things by *external* valuers.

16. The Long Line is the line of ordinal numbers defined by the *Biggest Set Theory*. The Biggest Set Theory is von Neumann–Gödel–Bernays set theory (NGB) plus axioms for all consistently definable large cardinals. By itself, NGB is just standard Zermelo-Fraenkel-Choice (ZFC) set theory modified to allow proper classes (see Hamilton, 1982: chs. 4 & 6). For large cardinals, see Drake's 1974 classic.

17. Due to their divine perfections, the digital gods are appropriate objects of *veneration* and *reverence*. And you can be *grateful* to the gods for having brought you into being. If algorithms can be *praised* or *worshipped*, then so can the digital gods. Infinite digital gods are *sublime*; and some sublime gods are *numinous*.

18. Monotheists may object that digitalism is ontologically extravagant. After all, digitalism is an extreme type of polytheism – it posits a *proper class*

of gods. The monotheistic objection is that, since it is more parsimonious to posit one god rather than many, Ockham's razor cuts against digitalism and in favor of monotheism. Digitalists have two replies. The first reply is that Ockham's razor should be applied to types rather than tokens (Miller, 2001: sec. 4). If that is right, then positing an infinite plurality of gods is no worse than positing one. The second reply is that positing absolutely infinitely many little gods is in fact *not* more extravagant than positing one big absolutely infinite god. Positing infinitely many natural numbers is *not* more extravagant than positing the one infinite number $\aleph_0$. On the contrary, those two posits are equivalent in extravagance. Thus monotheists cannot use Ockham's razor to defeat digitalism.

19. As an illustration of earthly innovation, consider a blueprint for a bicycle and some ways to improve it. Here are some ways: make the ride more stable; make the ride smoother; make the steering less sensitive; make the drive train more reliable; make the tires less prone to flats; make the rims stronger; make the frame lighter. For each of these design problems, there are almost certainly many possible solutions.

20. Digital gods create their universes by running cosmic scripts, which consist of at least the laws of nature and initial conditions. The gods do not intervene in their universes, do not violate the laws of nature, do not perform miracles. Hence they do not affirmatively answer any prayers requesting intervention in nature on behalf of any person.

21. For more on divine evolution, see Steinhart (2013). The Algorithm for Divine Innovation is basically the Leibnizian Search for Software.

22. A universe u is minimally more perfect than some progression P if and only if it is the least perfect universe that is more perfect than every member of P. Thus u is minimally more perfect than P if and only if every x in P is less perfect than u and there is no universe v such that every x in P is less perfect than v and v is less perfect than u.

23. Against the combinatorial theory of possibility developed by Lewis (1986), Armstrong (1989: 29) argues that there is a hierarchy of possible universes. Digitalists mostly agree with Armstrong. However, while Armstrong starts his hierarchy with our universe, digitalists start their hierarchy with Universe Alpha (made by god Alpha).

24. All statements involving concrete possibility or necessity are interpreted relative to the cosmological hierarchy, which thus serves as the domain for modal quantification. It satisfies the postulates of Lewisian counterpart theory (1968, 1986). Digitalists interpret modal statements using Lewisian semantics. To put it informally, the *de dicto* modes look like this: for any proposition P, it is *necessary* that P if and only if P is true at every universe; it is *possible* that P if and only if P is true at some universe. The analysis of the *de re* modes uses counterpart theory. To put it informally, for any property p, for any thing x, x is *essentially* p if and only if x is p and every counterpart of x is p; and x is *accidentally* p if and only if x is p but some counterpart of x is not p.

25. Within any universe, time can be infinite into the past and future. Let the time in a universe be isomorphic to the unit interval [0, 1]. Let the present time be point 1/2. The infinitely past series is 1/4, 1/8, 1/16, and so on. The

infinite future series is 3/4, 7/8, 15/16, and so on. For each n, the unit interval $[n, n+1]$ is the time of a universe.

26. One metric for technical progress is the *Kardashev Scale*. Kardashev (1964) ranks societies on the basis of their energy consumption. He outlines three ranks. Rank I includes planetary societies that consume almost all the energy provided by their planetary resources (about 10^{19} ergs/sec). Our present earthly society is in Rank I. Rank II includes societies that consume almost all of the energy of their central star (about 10^{33} ergs/sec.). Such society may build *megastructures* – like Niven Rings or Dyson Spheres. Rank III includes societies able to harness almost all the energy of an entire galaxy (about 10^{44} erg/sec). As the indexes of universes in the cosmological hierarchy go up, the higher Kardashev ranks become more and more populated. Another metric for technical progress is the *Barrow Scale*. Barrow (1998: 133) describes seven types of human civilizations ranked by the smallness of things they can manipulate. The sizes of the objects are: (1) human; (2) genes; (3) molecules; (4) atoms; (5) atomic nuclei; (6) elementary particles; and (7) the structure of space-time itself. As the indexes of universes go up, the higher Barrow ranks become more and more populated.

VII Revision

1. Let P be a progression of universes defined on some ordinal k. A progression of processes over P is a series f such that, for every h less than k, the process $f(h)$ comes from the universe P(h). Within any progression of processes, $f(n+1)$ is a successor of $f(n)$ for any n less than k and $f(\text{L})$ is a limit of $\langle f(h)|h < L\rangle$ for any limit L less than k.
2. For patterns of activity as software objects, see Dennett (1991). He distinguishes several levels of software: the physical, the design, and the intentional.
3. For any lovers x and y, the love of x and y is *possible across all successors* if and only if *at least one* way to improve any universe they inhabit contains some way to improve their love. The Optimistic Principle entails that all love is possible across all successors. Hence some successor universe will always realize that amorous possibility. For any lovers x and y, the love of x and y is *necessary across all successors* if and only if *every* way to improve any universe they inhabit contains some way to improve their love. The Optimistic Principle never entails that any love is necessary across all successors.
4. A progression of couples is *mated* if and only if there are progressions of lives x and y such that the n-th couple in the progression contains the n-th life of x and the n-th life of y. For any mated progression of loving couples, their love is *possible in the limit* if and only if at least one way to improve their separate progressions of lives binds the limits of those life-progressions into a limit loving couple. The Optimistic Principle entails that love is always possible in the limit. Hence every mated progression of loving couples converges to at least one limit loving couple. For any mated progression of loving couples, their love is *necessary in the limit* if and only if every way to improve their separate progressions of lives binds the limits of those life-progressions into a

limit loving couple. The Optimistic Principle does not entail that any love is necessary in the limit.

5. The tree of life for any life x is defined by four rules. For the initial number 0 on the Long Line, the initial level T(x, 0) contains exactly x. For every successor number $n+1$ on the Long Line, the successor generation T(x, $n+1$) contains every improvement of every life in T(x, n). For every limit number L on the Long Line, the limit generation T(x, L) contains every improvement of every progression of lives of x up to L. The proper class T(x) is the union of T(x, n) for all n on the Long Line. T(x) is the tree of life of x.

6. You might try to argue that there is a direct path from the concept of resurrection in the Bible to the concept of revision. The Biblical thinks of resurrection as the *revival* of a corpse. The revival conception changes quickly to the Medieval concept of resurrection as the *reassembly* of the parts of the original body. With Hick (1976: ch. 15), the reassembly concept becomes *replication*. Replication changes into resurrection as *re-creation* (see Shorter, 1962: 81–4; Sutherland, 1964: 386; Hick, 1976: 465; Forrest, 1995: 58; Kundera, 1999: part 5, sec. 16; Steinhart, 2008). However, this evolution of the concept of resurrection changes it into a new concept: it becomes *rebirth*.

7. Although reincarnation is often thought of as an Eastern doctrine, it is common among the ancient Greeks. Plato describes reincarnation in the Myth of Er (*Republic*, 614b–621d). The Myth of Er differs from revision in many obvious ways. Reincarnation is also endorsed by Neoplatonists like Plotinus (*Enneads*, 3.2–4).

8. Counterpart theory supports branching revision. Consider Dilley's (1983) example in which Hick is to be revised both as a plumber and as a lawyer. For this to be true of Hick now, he must be able to truly say "I will be a plumber and I will be a lawyer." Counterpart theory says that the statement is true if and only if Hick has a counterpart Hick-1 who is a plumber and not a lawyer and Hick has a counterpart Hick-2 who is a lawyer but not a plumber. Since branching revision entails that Hick has both counterparts, he can truly make the statement.

9. Your life-script need not be determined in every detail. The script may at places be very specific, at other places it may be a vague roadmap. It may specify that you will go from point A to point B, but not how you will get there. It may allow for considerable nondeterminism (or indeterminism). Of course, nondeterminism is not libertarian free will. Digitalism handles nondeterminism by allowing time to branch. When time branches, things fission. Branching time is not considered here.

VIII Superhuman Bodies

1. Any human body which is superior to an earthly human body is *glorified*. Many writers have discussed glorified bodies. Olaf Stapledon, in his brilliant novel *First and Last Men,* posits 18 ranks of glorified bodies. Kurzweil (2005: 300–12) posits two ranks of glorified bodies. Many transhumanist writers discuss glorified bodies.

2. Your essence contains your earthly soul – the body-program encoded in your present earthly genetics. But your essence contains other souls – it contains

body-programs with better genetics. Two rules define your essence so far: (1) your earthly soul is in your essence; (2) for every soul P, if P is in your essence, then every optimization of P is in your essence. There are four main optimizations: (1) the correction of personal genetic defects; (2) the correction of species-level genetic defects; (3) the upgrading of all genetic circuits to the best human levels; and (4) the generalization of genetic circuitry.

3. Kurzweil (2005: 310) suggests that future humans may develop bodies capable of shape-shifting (metamorphosis). Preservation of the human body-plan is essential to being human. If you could change into a wolf (like a were-wolf) or into a bird or a fish, why would you still be human in any sense? You wouldn't.

4. Your essence contains many souls. It contains many body-programs. Three rules define your essence so far: (1) your earthly soul is in your essence; (2) for every soul P, if P is in your essence, then every optimization of P is in your essence; and (3) for every soul P, if P is in your essence, then every idealization of P is in your essence.

5. The vocal systems of idealized humans are biologically universal. They can make any sounds any carbon-based organism can make. An idealized brain is smart enough to understand all animal signaling systems. Hence the idealized vocal system can be used to communicate via sound with any other organism.

6. Your essence contains many souls. It contains many body-programs. Four rules define your essence so far: (1) your earthly soul is in your essence; (2) every optimization of any soul in your essence is also in your essence; (3) every idealization of every soul in your essence is also in your essence; and (4) every extension of every soul in your essence is also in your essence.

IX Infinite Bodies

1. Every particle of excelsium is an infinitary machine. It has infinitely many possible inputs, states, and outputs. And it has infinitely many if-then rules to define its changes. An infinitary machine is a configuration (p, i, s, o). The item p is an infinitary program; it is the soul of the machine. The item i is the input to the machine, s is its state, and o is its output. An infinitary program is a tuple (I, S, O, F, G) in which I, S, and O are infinite. An infinitary machine can accelerate to and through limits. It is computationally equivalent to an accelerating Turing machine (see Copeland, 1998a, 1998b).

2. For discussions and many examples of supertasks, see Earman & Norton (1996); Koetsier & Allis (1997); and Steinhart (2003).

3. Your essence contains many souls. Five rules define your essence so far: (1) your earthly soul is in your essence; (2) every optimization of any soul in your essence is also in your essence; (3) every idealization of any soul in your essence is also in your essence; (4) any extension of any soul in your essence is also in your essence; and (5) for any progression of increasingly extended souls, if your essence contains every soul in that progression, then your essence contains every limit of that progression.

4. Perhaps Helen made this string by mentally running an accelerating Turing machine to its limit. After all, a subtle brain can perform infinitary computations in finite time.

5. Supertasks are nested into deeper levels of Zeno fractions. The first level of Zeno fractions is the series 1/2, 3/4, 7/8, and so on. But every interval between any two Zeno fractions contains a deeper level of Zeno fractions. Between 0 and 1/2, the next level of Zeno fractions is the series that starts out as 1/4, 3/8, 7/16, and 15/32.

6. An infinitely long series of symbols drawn from some alphabet is a super-sentence in some super-language. An infinitely long bit string is one example of a super-sentence. The language of thought of any super-brain is some super-language and its thoughts are super-sentences. Super-languages include infinitary versions of the predicate calculus (Karp, 1964). Any Borges Book contains infinitely many super-sentences written in some super-language. For any cardinals κ and λ, if at least one of κ or λ is greater than ω, then there is a super-language L(κ, λ). Every super-language is surpassed by endlessly many more expressive super-languages.

7. A subtle body can compute the halting function for all classical Turing machines. It can build an infinite table that lists the recursive total functions from the positive integers to {0,1} and it can do anti-diagonalization on that table to obtain a nonrecursive function. It can ascend the hierarchy of Turing degrees.

8. For infinite games, Gale & Stewart (1953) is the classical text. Freiling (1984) gives many examples. Jech (1984), Ciesielski & Laver (1990), and Scheepers (1993) describe infinite games involving infinite memory and skill.

X Nature

1. Sets serve as purely mathematical models of Platonic forms (of god-forms, universe-forms, process-forms, and body-forms). But the equivalence class of models of any form can serve either as a proxy for that form or as the form itself. Sets can be coded as bit strings (via Ackermann's coding, or von Neumann's axioms). Since concepts like logical depth (that is, logical density) apply to bit strings, sets have logical depths. They are more or less complex; they have abstract degrees of intrinsic value. The *landscape function* maps every set in the proper class of sets onto some ordinal degree of intrinsic value. The epic of theology, and all the other epics, defines optimization algorithms that search for increasingly valuable set-theoretic structures in this abstract landscape. As those structures are found, they are concretely instantiated according to the rules of those epics.

2. Axiarchism pictures reality as ultimately ruled by value; extreme axiarchism states that this ultimate rule of value is principled or lawful rather than the result of divine personal intentions (Leslie, 1970: 286; 1979: 6). Despite his theism, Rescher offers an extreme axiarchism (1984, 2000). Digital axiarchism is also extreme.

3. Rescher states his *Principle of Axiology* like this: "among otherwise equally possible law-arrangements, that one is (or tends to be) realized which is maximally value-enhancing" (1984: 43). Rescher later states his *law of optimality* like this: "whatever possibility is for the best is *ipso facto* the possibility that is actualized" (2000: 815). Although ideas like these may be regarded as the

very distant ancestors of the digitalist Axiarchic Principle, digital ontology differs radically from Rescherian ontology.

4. On the basis of the long Platonic tradition in Western theology, Leslie points out that some non-traditional theists may want to use "God" to refer to creatively effective goodness (1970: 297; 1979: 1.1, 1.8; 1989: ch. 8; 2001: 179–82). Thus the Axiarchic Principle would be God. But Leslie dismisses this identification as merely verbal (2001: 185–6). And if the Axiarchic Principle is God, then God is not a god. It would be best to avoid the term "God" altogether, and to develop new and more appropriate vocabulary.

5. The good itself is not any Platonic sun above all things; on the contrary, the good itself is underneath everything. For digitalists, the good itself is the *seed* buried in the abstract background. The *unfolding* of all the meaning *enfolded* in this abstract seed is its eruption into concreteness. This eruption, this self-revelation, this unconcealedness, this *aletheia*, brings the Great Tree of gods into being. The Great Tree is the ramification of the good itself. Plotinus describes the world as a kind of tree whose buried root is the One (*Enneads*, 3.3.7, 3.8.10). He also describes the One as the hidden spring from which all being flows (*Enneads*, 3.8.10). The terms "enfolding" and "unfolding" come from Nicholas of Cusa. Of course, all that is said in this note is merely metaphorical.

6. Rescher and Leslie both run arguments from their abstract axiarchic principles to the existence of concrete things (Rescher, 1984: ch. 1; Leslie, 1989: chs. 6–8).

7. Rescher insists that intrinsic value is the only type of value that plays a role in axiarchic explanations (1984: ch. 2.3). He explicitly rejects anthropic or personal standards of value in favor of impersonal ontological and cosmological standards. The axiarchic values include "simplicity, harmony, systemic elegance, uniformity, etc." (1984: 49). For digitalists, all these values are incorporated into logical depth (density).

8. One version of *pantheism* asserts that (1) some unity contains all things and (2) the maximally-inclusive unity is divine (see MacIntyre, 1967: 34; Levine, 1994; Oppy, 1997: 320; Steinhart, 2004). Digitalists can thus be pantheistic about the plenum: the plenum is a maximally-inclusive unity which may be divine. Some digitalists may therefore want to use "God" to refer to the plenum. However, since the plenum is not a god, any such reference leads to the paradoxical conclusion that God is not a god. To avoid that paradox, it seems best to avoid the term "God". The plenum is not God. It would be better for pantheists to come up with their own distinctive conceptual vocabulary.

9. Plotinus was an early axiarchist who argued that everything derives from the power of the One, which is purely good. But then how to explain evil? Plotinus says it emerges from the conflicts among goods (*Enneads*, 4.4.32). And yet, since nature is animated by conflict, even this conflict is good (*Enneads*, 2.3.16). For Plotinus, all evil is in the parts but the whole of reality is good (*Enneads*, 3.2.3, 3.2.11, 3.2.17, 4.4.32). Digitalists can agree with these points without adopting any further Plotinian metaphysics.

10. To say that everything that ought to be true will be true means that for every concrete thing x, for every property p, if $p(x)$ ought to be true, then there

exists some future counterpart y of x such that $p(y)$ is true. Of course, y is a revision of x.

11. Since Leslie posits an infinite plurality of divine minds (2001), he is a polytheist. But his divine minds are all Anselmian: they are all maximally perfect. Digitalists deny that any maximally perfect minds exist – all minds are surpassable. Nevertheless, Leslian polytheism helps to set the stage for digital polytheism.

12. One version of logical depth (density) is known as *parallel depth*. Machta writes that parallel depth "can only become large for systems with embedded computation" (2011: 037111–1); and that "depth is sensitive to embedded computation and can only be large for systems that carry out computationally complex information processing" (037111–6). Hence as the depths (densities) of computations grow larger and larger, they begin to support ever higher stacks of internal virtual computations.

References

Aaronson, S. (2008) 'The limits of quantum computing', *Scientific American 298* (3), 62–9.

Abram, C. (2012) *Facebook for Dummies*. Indianapolis: John Wiley & Sons.

Adami, C., Ofria, C., & Collier, T. (2000) 'Evolution of biological complexity', *Proceedings of the National Academy of Sciences 97* (9), 4463–8.

Agar, N. (2001) *Life's Intrinsic Value: Science, Ethics, and Nature*. New York: Columbia University Press.

Alliksaar, M. (2001) 'Metabolism in A-Life: Reply to Boden', *British Journal for the Philosophy of Science 52* (1), 131–5.

Alon, U. (2007) *Introduction to Systems Biology: Design Principles of Biological Circuits*. Boca Raton, FL: Chapman & Hall.

Alvarez, R. (2000) 'Genetic variation in the renin-angiotensin system and athletic performance', *European Journal of Applied Physiology 82*, 117–20.

Antunes, L., Fortnow, L., Melkebeek, D., & Vinodch, N. (2006) 'Computational depth: Concept and applications', *Theoretical Computer Science 354* (3), 391–404.

Antunes, L., Matos, A., Souto, A., & Vitanyi, P. (2009) 'Depth as randomness deficiency', *Theory of Computing Systems 45*, 724–39.

Armstrong, D. (1989) *A Combinatorial Theory of Possibility*. New York: Cambridge University Press.

Ay, N., Muller, M., & Szkola, A. (2010) 'Effective complexity and its relation to logical depth', *IEEE Transactions on Information Theory 56* (9), 4593–607.

Bainbridge, W. S. (2003) 'Massive questionnaires for personality capture', *Social Science Computer Review 21* (3), 267–80.

Bainbridge, W. S. (2010) *The Warcraft Civilization: Social Science in a Virtual World*. Cambridge, MA: The MIT Press.

Baker, L. R. (2000) *Persons and Bodies: A Constitution View*. New York: Cambridge University Press.

Balashov, Y. (2000a) 'Persistence and space-time: Philosophical lessons of the pole and the barn', *The Monist 83*, 321–40.

Balashov, Y. (2000b) 'Relativity and persistence', *Philosophy of Science 67*, S549–62.

Barabasi, A.-L. & Oltvai, Z. (2004) 'Network biology: Understanding the cell's functional organization', *Nature Reviews Genetics 5*, 101–13.

Barrett, J. & Aitken, W. (2010) 'A note on the physical possibility of ordinal computation', *British Journal for the Philosophy of Science 61*(4), 867–74.

Barrow, J. (1998) *Impossibility: The Limits of Science and the Science of Limits*. New York: Oxford University Press.

Barrow, J. & Tipler, F. (1986) *The Anthropic Cosmological Principle*. New York: Oxford University Press.

Basalla, G. (1988) *The Evolution of Technology*. New York: Cambridge University Press.

Bassingthwaighte, J., Liebovitch, L., & West, B. (1994) *Fractal Physiology*. New York: Oxford University Press.

Bays, C. (2005) 'A note on the game of life in hexagonal and pentagonal tessellations', *Complex Systems 15*, 245–52.

Bays, C. (2006) 'A note about the discovery of many new rules for the game of three-dimensional life', *Complex Systems 16*, 381–6.

Bedau, M. (1998) 'Philosophical content and method of artificial life', in T. Bynum & J. Moor (eds.) (1998) *The Digital Phoenix: How Computers Are Changing Philosophy*. Malden, MA: Basil Blackwell, 135–52.

Bekenstein, J. & Schiffer, M. (1990) 'Quantum limitations on the storage and transmission of information', *International Journal of Modern Physics 1*, 355–422.

Benjamini, E., Sunshine, G., & Leskowitz, S. (1996) *Immunology: A Short Course*. John Wiley & Sons, New York.

Bennett, C. (1985) 'Dissipation, information, computational complexity and the definition of organization', in D. Pines (ed.) (1988) *Emerging Syntheses in Science*. New York: Perseus Books, 297–313.

Bennett, C. (1988) 'Logical depth and physical complexity', in Herken, R. (ed.) (1988) *The Universal Turing Machine: A Half-Century Survey*. New York: Oxford University Press, 227–57.

Bennett, C. (1990) 'How to define complexity in physics, and why', in W. Zurek (ed.) (1990) *Complexity, Entropy, and the Physics of Information*. Reading, MA: Addison-Wesley, 137–48.

Bennett, J. (1984) *A Study of Spinoza's Ethics*. New York: Cambridge University Press.

Bhalla, U. & Iyengar, R. (1999) 'Emergent properties of networks of biological signaling pathways', *Science 283* (15 January), 381–7.

Blackledge, T. et al. (2009) 'Reconstructing web evolution and spider diversification in the molecular era', *Proceedings of the National Academy of Sciences 106* (13), 5229–34.

Blascovitch, J. & Bailenson, J. (2011) *Infinite Reality*. New York: Harper Collins.

Bleh, D., Calarco, T., & Montangero, S. (2012). 'Quantum game of life', *EPL (Europhysics Letters)*, 97 (2), 20012.

Blum, L., Cucker, F., Shub, M., & Smale, S. (1998) *Complexity and Real Computation*. New York: Springer-Verlag.

Blumenfeld, D. (2009) 'Living life over again', *Philosophy and Phenomenological Research 79* (2), 357–86.

Boahen, K. (2005) 'Neuromorphic microchips', *Scientific American 292* (5), 56–63.

Bonta, I. (2003) 'Schizophrenia, dissociative anaesthesia and near-death experience; three events meeting at the NMDA receptor', *Medical Hypotheses 62*, 23–8.

Borges, J. L. (1964) *Labyrinths*. New York: New Directions.

Bornholdt, S. (2005) 'Less is more in modeling large genetic networks', *Science 310*, 449–50.

Bornholdt, S. (2008) 'Boolean network models of cellular regulation: Prospects and limitations', *Journal of the Royal Society Interface 5*, S85–94.

Bostrom, N. (2003) 'Are you living in a computer simulation?', *Philosophical Quarterly 53* (211), 243–55.

Bostrom, N. (2005) 'The simulation argument: Reply to Weatherson', *Philosophical Quarterly 55* (218), 90–7.

Bower, J. (1988) *The Evolution of Complexity by Means of Natural Selection.* Princeton, NJ: Princeton University Press.

Bray, D. (1990) 'Intracellular signaling as a parallel distributed process', *Journal of Theoretical Biology 143*, 215–31.

Bray, D. (1995) 'Protein molecules as computational elements in living cells', *Nature 376* (July 27, 1995), 307–12.

Brey, X. (2008) 'Technological design as an evolutionary process', in Vermaas, P., Kroes, P., Light, A. & Moore, S. (eds.) *Philosophy and Design.* New York: Springer, 61–76.

Britton, W. & Bootzin, R. (2004) 'Near-death experiences and the temporal lobe', *Psychological Science 15* (4), 254–58.

Brockes, J. (1997) 'Amphibian limb regeneration: Rebuilding a complex structure', *Science 276* (5309), 81–7.

Broderick, D. (2001) *The Spike: How Our Lives Are Being Transformed by Rapidly Advancing Technologies.* New York: Tom Doherty Associates.

Brown, J. (1990) 'Interview with Edward Fredkin', *New Scientist* (14 July), 39.

Browning, D. (1988) 'Sameness through change and the coincidence of properties', *Philosophy and Phenomenological Research 49* (1), 103–21.

Brueckner, A. (2008) 'The simulation argument again', *Analysis 68* (299), 224–6.

Burks, A. (1973) 'Logic, computers, and men', *Proceedings and Addresses of the American Philosophical Association 46*, 39–57.

Bush, V. (1945) 'As we may think', *The Atlantic Monthly 176* (1) (July), 101–8.

Cameron, R. (2008) 'Turtles all the way down', *Philosophical Quarterly 58*, 1–14.

Cassell, J., Sullivan, J., Prevost, S., & Churchill, E. (eds.) (2000) *Embodied Conversational Agents.* Cambridge, MA: MIT Press.

Castronova, E. (2005) *Synthetic Worlds.* Chicago: University of Chicago Press.

Chalmers, D. (2005) 'The matrix as metaphysics', in C. Grau (ed.) (2005) *Philosophers Explore the Matrix.* New York: Oxford University Press, 132–76.

Chalmers, D. (2010) 'The singularity: A philosophical analysis', *Journal of Consciousness Studies 17*, 7–65.

Churchland, P. (1985) *Matter and Consciousness.* Cambridge, MA: The MIT Press.

Ciesielski, K. & Laver, R. (1990) 'A game of D. Gale in which one of the players has limited memory', *Periodica Mathematica Hungarica 22*, 153–8.

Clark, P. & Read, S. (1984) 'Hypertasks', *Synthese 61*, 387–90.

Clay, D. & Purvis, A. (1999) *Four Island Utopias.* Newburyport, MA: Focus Publishing.

Coenen, C. (2007) 'Utopian aspects of the debate on converging technologies', in G. Banse et al. (eds.) (2007) *Assessing Societal Implications of Converging Technological Development.* Berlin: Edition Sigma, 141–72.

Collias, N. (1964) 'The evolution of nests and nest-building in birds', *American Zoologist 4* (2), 175–90.

Collias, N. (1997) 'On the origin and evolution of nest building by passerine birds', *The Condor 99* (2), 253–70.

Colyvan, M. (2001) *The Indispensability of Mathematics.* New York: Oxford University Press.

Colyvan, M., Garfield, J., & Priest, G. (2005) 'Problems with the argument from fine tuning', *Synthese 145*, 325–38.

Conrad, E. & Tyson, J. (2006) 'Modeling molecular interaction networks with nonlinear ordinary differential equations', in Z. Szallasi et al. (eds.) (2006), 97–124.

Conway, J. (2001) *On Numbers and Games*. Second Edition. Natick, MA: A. K. Peters.

Copeland, B. J. (1998a) 'Super-Turing machines', *Complexity 4* (1), 30–2.

Copeland, B. J. (1998b) 'Even Turing machines can compute uncomputable functions', in C. Calude, J. Casti, and M. Dinneen (eds.), *Unconventional Models of Computation*. New York: Springer-Verlag, 150–64.

Cowan, J. (1974) 'The paradox of omniscience revisited', *Canadian Journal of Philosophy 3* (3), 435–45.

Crutchfield, J. & Mitchell, M. (1995) 'The evolution of emergent computation', *Proceedings of the National Academy of Sciences 92* (23), 10742–6.

Cuthbertson, R., Paton, R., & Holcombe, M. (1996) *Computation in Cellular and Molecular Biological Systems*. River Edge, NJ: World Scientific.

Davis, S. (1992) 'Hierarchical causes in the cosmological argument', *International Journal for the Philosophy of Religion 31* (1), 13–27.

Davison, S. (2010) *On the Intrinsic Value of Everything*. New York: Continuum.

Dawkins, R. (1986) *The Blind Watchmaker*. New York: W. W. Norton.

Dawkins, R. (1996) *Climbing Mount Improbable*. New York: W. W. Norton.

Dawkins, R. (2008) *The God Delusion*. New York: Houghton-Mifflin.

De Ridder, D. et al. (2007) 'Visualizing out-of-body experiences in the brain', *New England Journal of Medicine 357*, 1829–33.

Dennett, D. (1978/1981) *Brainstorms*. Cambridge, MA: The MIT Press.

Dennett, D. (1991) 'Real patterns', *The Journal of Philosophy 88* (1), 27–51.

Dennett, D. (1993) *Consciousness Explained*. New York: Penguin.

Dennett, D. (1995) *Darwin's Dangerous Idea: Evolution and the Meanings of Life*. New York: Simon & Schuster.

Deutsch, D. (1985) 'Quantum theory, the Church-Turing principle and the universal quantum computer', *Proceedings of the Royal Society*, Series A, 400, 97–117.

Diamond, J. (1989) 'How cats survive falls from New York skyscrapers', *Natural History* (August), 20–6.

Dilley, F. (1983) 'Resurrection and the "Replica objection"', *Religious Studies 19*, 459–74.

Djurfeldt, M. et al. (2008) 'Brain-scale simulation of the neocortex on the IBM Blue Gene/L supercomputer', *IBM Journal of Research and Development 51* (1/2), 31–41.

Doore, G. (1980) 'The argument from design: Some better reasons for agreeing with Hume', *Religious Studies 16* (2), 145–61.

Douglas, T. (2008) 'Moral enhancement', *Journal of Applied Philosophy 25* (3), 228–45.

Dowe, P. (1992) Wesley Salmon's process theory of causality and the conserved quantity theory. *Philosophy of Science 62*, 321–333.

Dozois, G. (ed.) (1991) *Writing Science Fiction and Fantasy*. New York: St. Martin's Press.

Drake, F. (1974) *Set Theory: An Introduction to Large Cardinals*. New York: American Elsevier.

Dworkin, R. (1993) *Life's Dominion*. New York: Knopf.

Dyson, G. (1997) *Darwin among the Machines: The Evolution of Global Intelligence*. Reading, MA: Perseus Books.

Dyson, G. (2012) *Turing's Cathedral: The Origins of the Digital Universe*. New York: Vintage Press.

Earman, J. & Norton, J. (1996) 'Infinite pains: The trouble with supertasks', in A. Morton & S. Stich (eds.) *Benacerraf and His Critics*. Cambridge, MA: Blackwell, 231–62.

Ehrsson, H. (2007) 'The experimental induction of out-of-body experiences', *Science 317* (24 August), 1048.

Elliot, R. (1992) 'Intrinsic value, environmental obligation and naturalness', *The Monist 75* (2), 138–60.

Emery, N. (2006) 'Cognitive ornithology: The evolution of avian intelligence', *Philosophical Transactions of the Royal Society B 361*, 23–43.

Eppstein, D. (2010) 'Growth and decay in life-like cellular automata', in A. Adamatzky (ed.) (2010) *Game of Life Cellular Automata*. New York: Springer-Verlag, 71–98.

Evans, K. M. (2003) 'Larger than life: threshold-range scaling of life's coherent structures', *Physica D 183*, 45–67.

Faust, H. (2008) 'Should we select for moral enhancement?', *Theoretical Medicine and Bioethics 29*, 397–416.

Fisher, J. & Henzinger, T. (2007) 'Executable cell biology', *Nature Biotechnology 25* (11), 1239–49.

Flitney, A. & Abbot, D. (2005) 'A semi-quantum version of the game of life', *Advances in Dynamic Games 7*, 667–79.

Forrest, P. (1995) *God without the Supernatural: A Defense of Scientific Theism*. Ithaca, NY: Cornell University Press.

Forrest, P. (2007) *Developmental Theism: From Pure Will to Unbounded Love*. New York: Oxford University Press.

Frankfurt, H. (1971) 'Freedom of the will and the concept of a person', *Journal of Philosophy 68* (1), 5–20.

Franklin, B. (1771) 'The autobiography', in A. Houston (ed.) (2004) *The Autobiography and other Writings on Politics, Economics, and Virtue*. New York: Cambridge University Press, 1–142.

Fredkin, E. (1991) 'Digital mechanics: An informational process based on reversible universal cellular automata', in Gutowitz, H. (1991) (ed.) *Cellular Automata: Theory and Experiment*. Cambridge, MA: MIT Press, 254–70.

Fredkin, E. (1992) 'A new cosmogony', in D. Matzke (ed.) (1993) *Workshop on Physics and Computation*. Los Alamitos, CA: IEEE Computer Society Press, 116–21.

Fredkin, E. (2003) 'An introduction to digital philosophy', *International Journal of Theoretical Physics 42* (2), 189–247.

Fredkin, E., Landauer, R., & Toffoli, T. (eds.) (1982) 'Physics of computation (Conference Proceedings)', *International Journal of Theoretical Physics*. Part I: Vol. 21, Nos. 3 & 4 (April 1982); Part II: Vol. 21, Nos. 6 & 7 (June 1982); Part III: Vol. 21, No. 12 (December 1982).

Freiling, C. (1984) 'Banach games', *Journal of Symbolic Logic 49* (2), 343–75.

Gale, D. & Stewart, F. M. (1953) 'Infinite games with perfect information', in H. Kuhn & A. Tucker (eds.), *Contributions to the Theory of Games Vol. 2*. Annals of Mathematical Studies 28. Princeton, NJ: Princeton University Press, 245–66.

Gara, A. et al. (2005) 'Overview of the Blue Gene/L system architecture', *IBM Journal of Research and Development 49* (2/3), 195–212.

Gardner, M. (1970) 'The fantastic combinations of John Conway's new solitaire game "life"', *Scientific American 223*, 120–3.

Gattringer, C. & Lang, C. (2009) *Quantum Chromodynamics on the Lattice.* New York: Springer.

Gell-Mann, M. (1995) 'What is complexity?', *Complexity 1* (1), 16–9.

Gemmell, J., et al. (2002) 'MyLifeBits: fulfilling the memex vision', *ACM Multimedia '02*, December 1–6, 2002, Juan-les-Pins, France, 235–8.

Gemmell, J., Lueder, R., & Bell, G. (2003) 'The MyLifeBits lifetime store', *ACM SIGMM 2003 Workshop on Experiential Telepresence* (ETP 2003), November 7, 2003, Berkeley, CA.

Gershon, M. (1998) *The Second Brain.* New York: Harper-Collins.

Gibson, W. (1984) *Neuromancer.* New York: ACE Books.

Gibson, W. (1988) *Mona Lisa Overdrive.* New York: Bantam Books.

Gillett, S. & Bova, B. (2001) *World Building: A Writer's Guide to Constructing Star Systems and Life-Supporting Planets.* Cincinnati, OH: Writers Digest Books.

Giunti, M. (1997) *Computation, Dynamics, and Cognition.* New York: Oxford University Press.

Goertzel, B. & Bugaj, V. S. (2006) *The Path to Posthumanity: 21st Century Technology and its Radical Implications for Mind, Society, and Reality.* Bethesda, MD: Academica Press.

Goldberger, A., Rigney, D., & West, B. (1990) 'Chaos and fractals in human physiology', *Scientific American 262* (2), 42–9.

Good, I. (1965) 'Speculations concerning the first ultraintelligent machine', in F. Alt & M. Rubinoff (eds.) *Advances in Computers 6.* New York: Academic Press, 31–88.

Grim, P. (1988) 'Logic and limits of knowledge and truth', *Nous 22*, 341–67.

Guirao, A., Cox, I., & Williams, D.R. (2002) 'Method for optimizing the benefit of correcting the eye's higher-order aberrations in the presence of decentrations', *Journal of the Optical Society of America A 19* (1), 126–8.

Hales, S. & Johnson, T. (2003) 'Endurantism, perdurantism and special relativity', *Philosophical Quarterly 53* (213), 524–39.

Hallett, M. (1988) *Cantorian Set Theory and Limitation of Size.* New York: Oxford University Press.

Hamilton, A. (1982) *Numbers, Sets, and Axioms: The Apparatus of Mathematics.* New York: Cambridge University Press.

Hamkins, J. (2002) 'Infinite time Turing machines', *Minds and Machines 12* (4), 521–39.

Hamkins, J. & Lewis, A. (2000) 'Infinite time Turing machines', *Journal of Symbolic Logic 65* (2), 567–604.

Hansell, M. (2005) *Animal Architecture.* New York: Oxford University Press.

Hanson, R. (2001) 'How to live in a simulation', *Journal of Evolution and Technology 7*, 1. Online at <http://jetpress.org/volume7/simulation.html>.

Harris, S. (2008) *Letter to a Christian Nation.* New York: Vintage Books.

Harrison, C. (1974) 'Totalities and the logic of first cause arguments', *Philosophy and Phenomenological Research 35* (1), 1–19.

Hartwell, L., Hopfield, J., Leibler, S., & Murray, A. (1999) 'From molecular to modular cell biology', *Nature 402 Supplement*, C47–52.

Harvell, B. (2012) *Teach Yourself Visually: Facebook.* Indianapolis: John Wiley & Sons.

Haslanger, S. & Kurtz, R. M. (2006) *Persistence: Contemporary Readings.* Cambridge, MA: MIT Press.

Hawley, K. (2001) *How Things Persist.* New York: Oxford University Press.

Held, L. (2009) *Quirks of Human Anatomy: An Evo-Devo Look at the Human Body.* New York: Cambridge University Press.

Helikar, T., Konvalina, J., Heidel, J., & Rogers, J. (2008) 'Emergent decision-making in biological signal transduction networks', *Proceedings of the National Academy of Sciences 105* (6), 1913–18.

Heller, M. (1990) *The Ontology of Physical Objects.* New York: Cambridge University Press.

Hellingwerf, K. (2005) 'Bacterial observations: A rudimentary form of intelligence?', *Trends in Microbiology 13* (4), 152–8.

Helms, V. (2008) *Principles of Computational Cell Biology: From Protein Complexes to Cellular Networks.* Weinheim, Germany: Wiley-VCH.

Hick, J. (1976) *Death and Eternal Life.* New York: Harper & Row.

Hillis, D. & Tucker, L. (1993) 'The CM-5 Connection Machine: A scalable supercomputer', *Communications of the ACM 36* (11), 31–40.

Hobbes, T. (1660/1962) *Leviathan.* New York, NY: Collier Books.

Honkavaara, J. et al. (2002) 'Ultraviolet vision and foraging in terrestrial vertebrates', *OIKOS: A Journal of Ecology 98* (3), 505–11.

Hopcroft, J. (1984) 'Turing machines', *Scientific American 250* (5), 86–107.

Hudson, H. (2001) *A Materialist Metaphysics of the Human Person.* Ithaca, NY: Cornell University Press.

Hume, D. (1779/1990) *Dialogues Concerning Natural Religion.* New York: Penguin.

Hunter, P. & Borg, T. (2003) 'Integration from proteins to organs: the Physiome Project', *Nature Reviews Molecular Cell Biology 4*, 237–43.

Izhikevich, E. & Edelman, G. (2008) 'Large-scale model of mammalian thalamocortical systems', *Proceedings of the National Academy of Sciences 105* (9), 3593–8.

Jansen, K. (1997) 'The ketamine model of the near-death experience', *Journal of Near-Death Studies 16* (1), 5–26.

Jech, T. (1984) 'More game-theoretic properties of boolean algebras', *Annals of Pure and Applied Logic 26*, 11–29.

Kane, R. (1986) 'Principles of reason', *Erkenntnis 24*, 115–136.

Kardashev, N. (1964) 'Transmission of information by extraterrestrial civilizations', *Soviet Astronomy – AJ 8* (2), 217–21.

Karp, C. (1964) *Languages with Expressions of Infinite Length.* Amsterdam: North-Holland Publishing Company.

Karr, J. et al. (2012) 'A whole-cell computational model predicts phenotype from genotype', *Cell 150*, 389–401.

Kaston, B. (1964) 'The evolution of spider webs', *American Zoologist 4* (2), 191–207.

Keller, A., et al. (2012) 'New insights into the Tyrolean Iceman's origin and phenotype as inferred by whole-genome sequencing', *Nature Communications 3* (698), 1–9.

Kempf, K. (1961) *Electronic Computers within the Ordnance Corps.* Aberdeen Proving Ground, Maryland: U.S. Army Ordnance Corps.

King, J. et al. (2009) 'A component-based extension framework for large-scale parallel simulations in NEURON', *Frontiers in Neuroscience 3* (10), 1–11.

Kirk, G. S. & Raven, J. E. (1957) *The Presocratic Philosophers.* New York: Cambridge University Press.

Kitcher, P. (1984) *The Nature of Mathematical Knowledge.* New York: Oxford University Press.

Koepke, P. (2005) 'Turing computations on ordinals', *Bulletin of Symbolic Logic 11*, 377–97.

Koepke, P. (2006a) 'Infinite time register machines', in A. Beckmann et al. (eds.) *Logical Approaches to Computational Barriers. Lecture Notes in Computer Science 3988*, 257–66.

Koepke, P. (2006b) 'Computing a model of set theory', in S. B. Cooper et al. (eds.) *New Computational Paradigms. Lecture Notes in Computer Science 3988*, 223–32.

Koepke, P. (2007) 'alpha-Recursion theory and ordinal computability', in I. Dimitriou, J. Hamkins, & P. Koepke (eds.) *BIWOC – Bonn International Workshop on Ordinal Computability.* Bonn Logic Reports, 48–55.

Koepke, P. & Siders, R. (2008) 'Register computations on ordinals', *Archive for Mathematical Logic 47*, 529–48.

Koetsier, T. & Allis, V. (1997) 'Assaying supertasks', *Logique et Analyse 159*, 291–313.

Koperski, J. (2005) 'Should we care about fine-tuning?', *British Journal for the Philosophy of Science 56*, 303–19.

Koppel, M. (1987) 'Complexity, depth, and sophistication', *Complex Systems 1*, 1087–91.

Kosinski, M., Stillwell, D., & Graepel, T. (2013) 'Private traits and attributes are predictable from digital records of human behavior', *Proceedings of the National Academy of Sciences 110* (15), 5802–5.

Kripke, S. (1980) *Naming and Necessity.* Cambridge, MA: Harvard University Press.

Kundera, M. (1999) *The Unbearable Lightness of Being.* Trans. M. Heim. New York: Harper Perennial Classics.

Kurzweil, R. (1999) *The Age of Spiritual Machines.* New York: Penguin.

Kurzweil, R. (2002) 'Replies to critics', in J. Richards (ed.) (2002) *Are We Spiritual Machines? Ray Kurzweil vs. The Critics of Strong AI.* Seattle, WA: The Discovery Institute.

Kurzweil, R. (2005) *The Singularity is Near: When Humans Transcend Biology.* New York: Viking.

La Mettrie, J. O. (1748/1912) *Man a Machine.* La Salle, IL: Open Court.

Langton, C. (1984) 'Self-reproduction in cellular automata', *Physica D 10*, 135–44.

Leibniz, G. W. (1697/1988) 'On the ultimate origination of the universe', in P. Schrecker & A. Schrecker (eds.) (1988) *Leibniz: Monadology and Other Essays.* New York: Macmillan Publishing, 84–94.

Lenggenhager, B. et al. (2007) 'Video ergo sum: Manipulating bodily self-consciousness', *Science 317* (24 August), 1096–9.

Leslie, J. (1970) 'The theory that the world exists because it should', *American Philosophical Quarterly 7* (4), 286–98.

Leslie, J. (1979) *Value and Existence.* Totowa, NJ: Rowman & Littlefield.

Leslie, J. (1989) *Universes.* New York: Routledge.

Leslie, J. (2001) *Infinite Minds: A Philosophical Cosmology.* New York: Oxford.

Levine, M. (1994) *Pantheism: A Non-Theistic Concept of Divinity.* New York: Routledge.

Lewis, D. (1968) 'Counterpart theory and quantified modal logic', *Journal of Philosophy 65*, 113–26.

Lewis, D. (1973) *Counterfactuals.* Cambridge, MA: Harvard University Press.

Lewis, D. (1986) *On the Plurality of Worlds*. Cambridge, MA: Blackwell.

Liang, J., Williams, D., & Miller, D. (1997) 'Supernormal vision and high resolution retinal imaging through adaptive optics', *Journal of the Optical Society of America A 14* (11), 2884–92.

Lipson, H. & Pollack, J. (2000) 'Automatic design and manufacture of robotic lifeforms', *Nature 406* (31 August 2000), 974–8.

Litt, A. et al. (2006) 'Is the brain a quantum computer?', *Cognitive Science 30*, 593–603.

Lloyd, S. (2002) 'Computational capacity of the universe', *Physical Review Letters 88* (23) (May), 237901–5.

Lloyd, S. & Pagels, H. (1988) 'Complexity as thermodynamic depth', *Annals of Physics 188*, 186–213.

Locke, J. (1690/1959) *An Essay Concerning Human Understanding*. New York: Dover Publications.

Lohn, J. & Reggia, J. (1995) 'Discovery of self-replicating structures using a genetic algorithm', *Proceedings of the 1995 IEEE International Conference on Evolutionary Computing*, Perth. Heidelberg: Springer, 678–83.

Lovejoy, A. (1936) *The Great Chain of Being*. Cambridge, MA: Harvard University Press.

Lowe, L. & Schaff, J. (2001) 'The Virtual Cell: A software environment for computational cell biology', *Trends in Biotechnology 19* (10), 401–6.

Machta, J. (2011) 'Natural complexity, computational complexity, and depth', *Chaos 21*, 0371111–8.

MacIntyre, A. (1967) 'Pantheism', in P. Edwards (ed.) *The Encyclopedia of Philosophy*. New York: Macmillan.

Mackay, D. (1997) 'Computer software and life after death', in P. Edwards (ed.) (1997) *Immortality*. Amherst, NY: Prometheus Books, 247–9.

Malaby, T. (2009) *Making Virtual Worlds: Linden Lab and Second Life*. Ithaca, NY: Cornell University Press.

Mandelbrot, B. (1978) *Fractals: Form, Chance, and Dimension*. San Francisco: W. H. Freeman.

Manson, N. (2000) 'There is no adequate definition of fine-tuned for life', *Inquiry 43* (3), 341–51.

Markram, H. (2006) 'The Blue Brain project', *Nature Reviews Neuroscience 7* (February), 153–60.

Marois, R. & Ivanoff, J. (2005) 'Capacity limits of information processing in the brain', *Trends in Cognitive Science 9* (6), 296–305.

Mayfield, J. (2007) 'Minimal history, a theory of plausible explanation', *Complexity 12* (4), 48–53.

Miller, R. (1992) 'Concern for counterparts', *Philosophical Papers 21* (2), 133–40.

Miller, R. (2001) 'Moderate modal realism', *Philosophia 28* (1–4), 3–38.

Moore, C. (1996) 'Recursion theory on the reals and continuous-time computation', *Theoretical Computer Science 162* (1), 23–44.

Moore, G. E. (1922) 'The conception of intrinsic value', in G. E. Moore (ed.) (1922) *Philosophical Studies*. London: Routledge & Kegan Paul, 253–75.

Moore, G. E. (1965) 'Cramming more components onto integrated circuits', *Electronics 38* (8) (19 April 1965), 114–17.

Moraru, I. et al. (2008) 'Virtual cell modeling and simulation software', *IET Systems Biology 2* (5), 352–62.

Moravec, H. (1988) *Mind Children: The Future of Robot and Human Intelligence.* Cambridge MA: Harvard University Press.

Moravec, H. (1992) 'Pigs in cyberspace', in R. Miller & M. Wolf (eds.) (1992) *Thinking Robots, An Aware Internet, and Cyberpunk Librarians: The 1992 LITA President's Program.* Chicago: Library and Information Technology Association, 15–21.

Moravec, H. (2000) *Robot: Mere Machine to Transcendent Mind.* New York: Oxford University Press.

Morris, M., Saez-Rodriguez, J., Sorger, P., & Lauffenburger, D. (2010) 'Logic-based models for the analysis of cell signaling networks', *Biochemistry 49,* 3216–24.

Morris, T. (1987) 'Perfect being theology', *Nous 21* (1), 19–30.

Narayan, S. (2009) 'Supercomputers: Past, present, and the future', *ACM Crossroads 15* (4), 7–10.

Neeman, I. (2004) *The Determinacy of Long Games.* New York: Walter de Gruyer.

Nehaniv, C. & Rhodes, J. (2000) 'The evolution and understanding of hierarchical complexity in biology from an algebraic perspective', *Artificial Life 6,* 45–67.

Nelson, K. et al. (2006) 'Does the arousal system contribute to near death experience?', *Neurology 66,* 1003–9.

Nietzsche, F. (1885/1978) *Thus Spake Zarathustra.* Trans. R. J. Hollingdale. New York: Penguin Books.

Noble, D. (2002) 'Modeling the heart – from genes to cells to the whole organ', *Science 295,* 1678–82.

Norman, D. (1992) *Turn Signals are the Facial Expressions of Automobiles.* New York: Addison-Wesley.

Nozick, R. (1989) *The Examined Life.* New York: Simon & Schuster.

Ochoa, G. & Osier, J. (1993) *The Writer's Guide to Creating a Science Fiction Universe.* Cincinnati, OH: Writers Digest Books.

Olshansky, S., Carnes, B., & Butler, R. (2001) 'If humans were built to last', *Scientific American 284* (3), 50–5.

Olson, E. (1997) *The Human Animal: Personal Identity without Psychology.* New York: Oxford University Press.

Oppy, G. (1997) 'Pantheism, quantification, and mereology', *The Monist 80* (2), 320–36.

Panger, M. et al. (2002) 'Older than Oldowan? Rethinking the emergence of hominin tool use', *Evolutionary Anthropology 11,* 235–45.

Parfit, D. (1971) 'On "The importance of self-identity"', *Journal of Philosophy 68* (20), 683–90.

Parfit, D. (1985) *Reasons and Persons.* New York: Oxford University Press.

Parker, E., Cahil, L., & McGaugh, J. (2006) 'A case of unusual autobigraphical remembering', *Neurocase 12,* 35–49.

Pearson, H. (2001) 'The regeneration gap', *Nature 414* (22 November), 388–90.

Pivato, M. (2007) 'RealLife: The continuum limit of Larger than Life cellular automata', *Theoretical Computer Science 372,* 46–68.

Polkinghorne, J. C. (1985) 'The scientific worldview and a destiny beyond death', in G. MacGregor (ed.) *Immortality and Human Destiny: A Variety of Views.* New York: Paragon House, 180–3.

Post, J. (1999) 'Naturalism', in R. Audi (ed.) *The Cambridge Dictionary of Philosophy.* New York: Cambridge University Press, 596–7.

Poundstone, W. (1985) *The Recursive Universe: Cosmic Complexity and the Limits of Scientific Knowledge.* Chicago: Contemporary Books Inc.

Powell, F. (2000) 'Respiration', in G. Causey Whittow (ed.) (2000), *Sturkie's Avian Physiology*. Fifth Edition. New York: Academic Press, 233–64.

Price, H. (1956) 'What kind of next world?', in P. Edwards (ed.) (1997) *Immortality*. Amherst, NY: Prometheus Books, 213–19.

Pring, L. (2005) 'Savant talent', *Developmental Medicine & Child Neurology 47*, 500–503.

Prusinkiewicz, P. & Lindenmayer, A. (1990) *The Algorithmic Beauty of Plants*. New York: Springer-Verlag.

Puccetti, R. (1969) 'Brain transplantation and personal identity', *Analysis 29* (3), 65–77.

Quine, W. (1950) 'Identity, ostension, and hypostasis', in S. Hales (ed.) (1999) *Metaphysics: Contemporary Readings*. Belmont, CA: Wadsworth, 461–8.

Quine, W. (1976) 'Wither physical objects?', *Boston Studies in the Philosophy of Science 39*, 497–504.

Quine, W. (1978) 'Facts of the matter', *Southwestern Journal of Philosophy 9* (2), 155–69.

Quine, W. (1990) 'Naturalism; Or, living within one's means', *Dialectica 49*, 251–61.

Rafler, S. (2011) 'Generalization of Conway's "Game of Life" to a continuous domain – SmoothLife'. Online at <arXiv: 1111.1567v2>.

Rahula, W. (1974) *What the Buddha Taught*. New York: Grove/Atlantic.

Reichenbach, B. (1978) 'Monism and the possibility of life after death', *Religious Studies 14* (1), 27–34.

Rendell, P. (2002) 'Turing universality of the game of life', in A. Adamatzky (ed.) (2002) *Collision-based Computation*. New York: Springer, 513–39.

Rescher, N. (1979) *Leibniz: An Introduction to his Philosophy*. Totowa, NJ: Rowman & Littlefield.

Rescher, N. (1984) *The Riddle of Existence: An Essay in Idealistic Metaphysics*. New York: University Press of America.

Rescher, N. (2000) 'Optimalism and axiological metaphysics', *The Review of Metaphysics 53* (4), 807–35.

Ripps, D. (2010) 'Using economy of means to evolve transition rules within 2D cellular automata', *Artificial Life 16* (2), 119–26.

Roberts, E., Magis, A., Ortiz, J., Baumeister, W., and Luthey-Schulten, Z. (2011) 'Noise contributions in an inducible genetic switch: A whole-cell simulation study', *PLoS Computational Biology 7* (3), 1–21.

Roberts, N., et al. (2009) 'A biological quarter-wave retarder with excellent achromaticity in the visible wavelength region', *Nature Photonics 3*, 641–4.

Robinson, A. (1996) *Non-Standard Analysis*. Revised Edition. Princeton, NJ: Princeton University Press.

Royce, J. (1899/1927) *The World and the Individual*. First Series. Supplementary Essay. New York: The Macmillan Company.

Rutherford, D. (1995) *Leibniz and the Rational Order of Nature*. New York: Cambridge University Press.

Rymaszewski, M. et al. (2007) *Second Life: The Official Guide*. Hoboken, NJ: John Wiley & Sons.

Sagan, C. (1995) 'Life', *Encyclopedia Britannica 22*, 985–1002.

Salen, K. & Zimmerman, E. (2003) *Rules of Play: Game Design Fundamentals*. Cambridge, MA: The MIT Press.

Salmon, W. C. (1966) 'Verifiablity and logic', in M. L. Diamond & T. V. Litzenburg (eds.) (1975) *The Logic of God: Theology and Verification*. Indianapolis, IN: Bobbs-Merrill, 456–79.

Salmon, W. (1984) *Scientific Explanation and the Causal Structure of the World*. Princeton, NJ: Princeton University Press.

Sandberg, A. (1999) 'The physics of information processing superobjects: Daily life among the Jupiter brains', *Journal of Evolution and Technology 5* (1), 1–34.

Sandberg, A. & Bostrom, N. (2008) *Whole brain emulation: A Roadmap*. Technical Report #2008–3, Future of Humanity Institute, Oxford University. Online at <www.fhi.ox.ac.uk/reports/2008–3.pdf>. Accessed 13 December 2010.

Sapin, E., Bailleux, O., Chabrier, J.-J., & Collet, P. (2004) 'A new universal cellular automaton discovered by evolutionary algorithms', *Lecture Notes in Computer Science Volume 3102*. New York: Springer, 175–87.

Sayama, H. (1999) 'A new structurally dissolvable self-reproducing loop evolving in a simple cellular automata space', *Artificial Life 5*, 343–65.

Scheepers, M. (1993) 'Variations on a game of Gale (I): Coding strategies', *Journal of Symbolic Logic 58* (3), 1035–43.

Schell, J. (2008) *The Art of Game Design: A Book of Lenses*. Burlington, MA: Elsevier Publishing.

Schmidhuber, J. (1997) 'A computer scientist's view of life, the universe, and everything', in C. Freksa (ed.) (1997) *Foundations of Computer Science: Potential – Theory – Cognition*. New York: Springer, 201–8.

Schmidhuber, J. (2007) 'Gödel machines: Fully self-referential optimal universal self-improvers', in B. Goertzel & C. Pennachin (eds.) (2007) *Artificial General Intelligence*. Berlin: Springer-Verlag, 199–226.

Schumaker, R., Walkup, K., & Beck, B. (2011) *Animal Tool Behavior: The Use and Manufacture of Tools by Animals*. Baltimore, MD: Johns Hopkins Press.

Scott, J. & Pawson, T. (2000) 'Cell communication: The inside story', *Scientific American 282* (6), 72–9.

Shafer, J. B., Konstan, J., & Riedl, J. (2001) 'E-commerce recommendation applications', *Data Mining and Knowledge Discovery 5*, 115–53.

Sharpeshkar, R. (2010) *Ultra Low Power Bioelectronics: Fundamentals, Biomedical Applications, and Bio-Inspired Systems*. New York: Cambridge University Press.

Shorter, J. (1962) 'More about bodily continuity and personal identity', *Analysis 22* (4), 79–85.

Sider, T. (1996) 'All the world's a stage', *Australasian Journal of Philosophy 74*, 433–53.

Sider, T. (2001) *Four-Dimensionalism: An Ontology of Persistence and Time*. New York: Oxford University Press.

Soule, M. (1985) 'What is conservation biology?', *BioScience 35* (11), 727–34.

Springel, V., et al. (2005) 'Simulations of the formation, evolution and clustering of galaxies and quasars', *Nature 435* (June 2), 629–36.

Stanford, C. (2003) *Upright: The Evolutionary Key to Becoming Human*. New York: Houghton Mifflin.

Steinhart, E. (1998) 'Digital metaphysics', in T. Bynum & J. Moor (eds.), *The Digital Phoenix*. New York: Basil Blackwell, 117–34.

Steinhart, E. (2001) 'Persons vs. brains: Biological intelligence in the human organism', *Biology and Philosophy 16* (1), 3–27.

Steinhart, E. (2002) 'Logically possible machines', *Minds and Machines 12* (2), 259–80.

Steinhart, E. (2003) 'Supermachines and superminds', *Minds and Machines 13*, 155–86.

Steinhart, E. (2004) 'Pantheism and current ontology', *Religious Studies 40* (1), 63–80.

Steinhart, E. (2007a) 'Survival as a digital ghost', *Minds and Machines 17*, 261–71.

Steinhart, E. (2007b) 'Infinitely complex machines', in A. Schuster (ed.) *Intelligent Computing Everywhere*. New York: Springer, 25–43.

Steinhart, E. (2008) 'The revision theory of resurrection', *Religious Studies 44* (1), 1–19.

Steinhart, E. (2009) *More Precisely: The Math You Need to do Philosophy*. Vancouver, BC: Broadview Press.

Steinhart, E. (2010) 'Theological implications of the simulation argument', *Ars Disputandi: The Online Journal for Philosophy of Religion 10*, 23–37.

Steinhart, E. (2012a) 'Ontology in the game of life', *Axiomathes 22* (3), 403–16.

Steinhart, E. (2012b) 'On the number of gods', *International Journal for the Philosophy of Religion 72* (2), 75–83.

Steinhart, E. (2013) 'On the plurality of gods', *Religious Studies 49* (3), 289–312.

Stoffregen, T. & Pittenger, J. (1995) 'Human echolocation as a basic form of perception and action', *Ecological Psychology 7* (3), 181–216.

Sutherland, S. (1964) 'Immortality and resurrection', *Religious Studies 3*, 377–89.

Suzuki, K. & Ikegami, T. (2006) 'Spatial-pattern-induced evolution of a self-replicating loop network', *Artificial Life 12*, 461–85.

Swan, M. (2012) 'Health 2050', *Journal of Personalized Medicine 2*, 93–118.

Szallasi, Z., Stelling, J., & Periwal, V. (eds.) (2006) *System Modeling in Cell Biology: From Concepts to Nuts and Bolts*. Cambridge, MA: The MIT Press.

Taft, R., Pheasant, M., & Mattick, J. (2007) 'The relationship between non-protein-coding DNA and eukaryotic complexity', *BioEssays 29* (3), 288–299.

Tegmark, M. (1998) 'Is "the Theory of Everything" merely the ultimate ensemble theory?', *Annals of Physics 270*, 1–51.

Tegmark, M. (2008) 'The mathematical universe', *Foundations of Physics 38*, 101–50.

Temkin, I. & Eldredge, N. (2007) 'Phylogenetics and material cultural evolution', *Current Anthropology 48* (1), 146–53.

Tipler, F. (1995) *The Physics of Immortality: Modern Cosmology, God and the Resurrection of the Dead*. New York: Anchor Books.

Tomita, M. (2001) 'Whole-cell simulation: A grand challenge of the twenty-first century', *Trends in Biotechnology 19* (6), 205–10.

Tomita, M. et al. (1999) 'E-CELL: Software environment for whole-cell simulation', *Bioinformatics 15* (1), 72–84.

Treffert, D. & Christensen, D. (2005) 'Inside the mind of a savant', *Scientific American 293* (6), 108–13.

Treffert, D. & Wallace, G. (2004) 'Islands of genius', *Scientific American Special Edition 14* (1), 14–23.

Turing, A. (1936) 'On computable numbers, with an application to the Entscheidungs-problem', *Proceedings of the London Mathematical Society 2* (42), 230–65.

Tyson, J., Chen, K., & Novak, B. (2003) 'Sniffers, buzzers, toggles, and blinkers: Dynamics of regulatory and signaling pathways in cells', *Current Opinion in Cell Biology 15*, 221–31.

van Inwagen, P. (1993) *Metaphysics*. San Francisco: Westview Press.

Viljanen, V. (2007) 'Field metaphysic, power, and individuation in Spinoza', *Canadian Journal of Philosophy 37* (3), 393–418.

Volrath, F. (1988) 'Untangling the spider's web', *Trends in Ecology and Evolution 3* (12), 331–35.

von Neumann, J. (1966) *The Theory of Self-Reproducing Automata*. Edited and finished by A. W. Burks. Urbana, IL: University of Illinois Press.

Walker, M., Dennis, T., & Kirshvink, J. (2002) 'The magnetic sense and its use in long-distance navigation by animals', *Current Opinion in Neurobiology 12*, 735–44.

Weatherson, B. (2003) 'Are you a sim?', *Philosophical Quarterly 53*, 425–31.

Wengert, R. (1971) 'The logic of essentially ordered causes', *Notre Dame Journal of Formal Logic 12* (4), 406–22.

Wiener, N. (1954) *The Human Use of Human Beings*. Garden City, NY: Doubleday Anchor Books.

Winkler, D. & Sheldon, F. (1993) 'Evolution of nest construction in swallows', *Proceedings of the National Academy of Sciences 90*, 5705–7.

Wolfram, S. (2002) *A New Kind of Science*. Champaign, IL: Wolfram Media.

Woods, D. et al. (2002) 'Insertion/deletion polymorphism of the angiotensin I-converting enzyme gene and arterial oxygen saturation at high altitude', *American Journal of Respiratory and Critical Care Medicine 166* (3), 362–6.

Wright, R. (1988) 'Did the universe just happen?', *The Atlantic Monthly* (April), 29–44.

Wuensche, A. (2004) 'Self-reproduction by glider collisions: The beehive rule', in J. Pollack et al. (eds.) (2004) *Artificial Life IX*. Cambridge, MA: MIT Press, 286–91.

Yan, L. et al. (2008) 'Modeling and simulation of hepatic drug disposition using a physiologically based, multi-agent in silico liver', *Pharmaceutical Research 25* (5), 1023–36.

Yoon, G. & Williams, D. (2002) 'Visual performance after correcting the monochromatic and chromatic aberrations of the eye', *Journal of the Optical Society of America A 19* (2), 266–75.

Zeilinger, A. (1999) 'A foundational principle for quantum mechanics', *Foundations of Physics 29* (4), 631–43.

Zenil, H. & Delahaye, J.-P. (2010) 'On the algorithmic nature of the world', in G. Dodig-Crnkovic & M. Burgin (eds.) *Information and Computation*. Singapore: World Scientific, 477–96.

Zuse, K. (1969). *Rechnender Raum*. Braunschweig, Germany: Vieweg & Sohn.

Zyskowski, K. & Prum, R. (1999) 'Phylogenetic analysis of the nest architecture of neotropical ovenbirds', *The Auk 116* (4), 891–991.

Index

GPSR Compliance
The European Union's (EU) General Product Safety Regulation (GPSR) is a set
of rules that requires consumer products to be safe and our obligations to
ensure this.

If you have any concerns about our products, you can contact us on

ProductSafety@springernature.com

In case Publisher is established outside the EU, the EU authorized
representative is:

Springer Nature Customer Service Center GmbH
Europaplatz 3
69115 Heidelberg, Germany